About Island Press

Since 1984, the nonprofit organization Island Press has been stimulating, shaping, and communicating ideas that are essential for solving environmental problems worldwide. With more than 1,000 titles in print and some 30 new releases each year, we are the nation's leading publisher on environmental issues. We identify innovative thinkers and emerging trends in the environmental field. We work with world-renowned experts and authors to develop cross-disciplinary solutions to environmental challenges.

Island Press designs and executes educational campaigns in conjunction with our authors to communicate their critical messages in print, in person, and online using the latest technologies, innovative programs, and the media. Our goal is to reach targeted audiences—scientists, policymakers, environmental advocates, urban planners, the media, and concerned citizens—with information that can be used to create the framework for long-term ecological health and human well-being.

Island Press gratefully acknowledges major support of our work by The Agua Fund, The Andrew W. Mellon Foundation, The Bobolink Foundation, The Curtis and Edith Munson Foundation, Forrest C. and Frances H. Lattner Foundation, The JPB Foundation, The Kresge Foundation, The Oram Foundation, Inc., The Overbrook Foundation, The S.D. Bechtel, Jr. Foundation, The Summit Charitable Foundation, Inc., and many other generous supporters.

The opinions expressed in this book are those of the author(s) and do not necessarily reflect the views of our supporters.

VITAL SIGNS

VITAL SIGNS

VOLUME 22

The Trends That Are Shaping Our Future

WORLDWATCH INSTITUTE

Michael Renner, *Project Director*

Robert Engelman   Max Lander
Jacqueline Espinal   Alvaro Lopez-Peña
Rabia Ferroukhi   Haibing Ma
Christoph von Friedeburg Michael Renner
Milena Gonzalez   Joel Stronberg
Gaelle Gourmelon   Yeneneh Terefe
Arslan Khalid   Xiangyu Wu
Mark Konold   Vincent Yi

Linda Starke, *Editor*
Lyle Rosbotham, *Designer*

ISLANDPRESS

Washington | Covelo | London

Contents

Acknowledgments

The articles in this book were first individually released on our companion online site, at vitalsigns.worldwatch.org, between April 2014 and March 2015.

Many individual and institutional funders provide the support without which this volume, as well as our other work, would not be possible. For their support during the past year of *Vital Signs*, *State of the World*, and a range of other reports and projects, we are deeply grateful to a wide range of organizations and individuals.

First of all, we thank our dedicated Board of Directors for their unflagging support and leadership: Ed Groark, Robert Charles Friese, John Robbins, L. Russell Bennett, Mike Biddle, Cathy Crain, Tom Crain, James Dehlsen, Edith Eddy, Christopher Flavin, Ping He, Nancy and Jerre Hitz, Izaak van Melle, Bo Normander, David Orr, and Richard Swanson, in addition to our Emeritus Directors, Øystein Dahle and Abderrahman Khene.

We also thank the many institutional funders whose support made Worldwatch's work possible over the past year. We are grateful to (in alphabetical order): the Ray C. Anderson Foundation; Asian Development Bank; Bieber Family Fund; Charles and Mary Bowers Living Trust; Carbon War Room Corporation; Caribbean Community; Climate and Development Knowledge Network; Cultural Vision Fund of the Orange County Community Foundation; Del Mar Global Trust; Doughty Hanson Charitable Foundation; Eaton Kenyon Fund of the Sacramento Region Community Foundation; Embassy of the Federal Republic of Germany to the United States; The Friese Family Fund; Garfield Foundation, Brian and Bina Garfield, Trustees; German Federal Ministry for the Environment, Nature Conservation and Nuclear Safety and the International Climate Initiative; William and Flora Hewlett Foundation with Population Reference Bureau; Hitz Foundation; Inter-American Development Bank; Steven Leuthold Family Foundation; National Renewable Energy Laboratory, U.S. Department of Energy; Renewable Energy Policy Network for the 21st Century; MAP Royalty, Inc., Natural Gas and Wind Energy Partnerships; Mom's Organic Market; Network for Good; Nutiva; Quixote Foundation, Inc.; Randles Family Living Trust; V. Kann Rasmussen Foundation; Estate of Aldean G. Rhyner; Serendipity Foundation; Shenandoah Foundation; Flora L. Thornton Foundation; Turner Foundation, Inc.; United Nations Foundation; United Way of Central New Mexico; Johanette Wallerstein Institute; Wallace Global Fund; Weeden Foundation Davies Fund; and World Bank–International Finance Corporation with CPCS Transcom Ltd.

For their financial contributions and in-kind donations, we thank Edith Borie, Charles and Angeliki Keil, John McBride, David and Mary Ellen Moir, Peter and Sara Ribbens, Peter Seidel, and five anonymous donors.

This edition was written by 18 researchers, including 4 non-Worldwatch authors. Befitting an organization with the word "World" in its name, they hail from countries around the globe: Robert Engelman, Jacqueline Espinal, Rabia Ferroukhi, Gary Gardner, Milena Gonzalez, Gaelle Gourmelon, Arslan Khalid, Mark Konold, Max Lander, Alvaro Lopez-Peña, Haibing Ma, Michael Renner, Joel Stronberg, Yeneneh Terefe, Christoph von Friedeburg, Xiangyu Wu, Vincent Yi, and Wanqing Zhou.

Editing this series since its inception in 1992, Linda Starke ensures consistency among the individual authors and skillfully enforces grammar rules and a clear writing style. Gaelle Gourmelon not only contributed two articles, but as part of her duties as Worldwatch's Marketing and Communications Manager, she polished authors' Word and Excel files and turned them into attractive PDFs for online posting, in addition to preparing the accompanying press releases. Thanks are also due to Lisa Mastny, who edits the press releases when individual *Vital Signs* articles are released online. Graphic designer Lyle Rosbotham is another indispensable veteran of the series, providing the layout for the book and selecting suitable images for the cover and section breaks.

No less important are the people who work hard to oversee our work, manage the office, and ensure that our work is funded. We thank in particular Ed Groark (Chairman of our Board and Acting Interim President), Barbara Fallin (Director of Finance and Administration), Mary Redfern (Director of Institutional Relations), and Development Associate Donald Minor, who also doubles as Administrative Assistant to the President.

We also express gratitude to our colleagues at Island Press, who share their ideas for book content, weigh in on other matters such as the cover illustration, and provide publicity for the book. In particular, we thank Charles C. Savitt, Emily Davis, Maureen Gately, Jaime Jennings, Julie Marshall, David Miller, and Sharis Simonian.

Michael Renner
Project Director
Worldwatch Institute
1400 16th Street, N.W.
Washington, DC 20036
vitalsigns.worldwatch.org

Introduction:
Consumption Choices Matter

Michael Renner

As participants in a global economy, we have gotten used to the idea that everything is somehow connected to everything else. Yet the logic of globalization regards people mostly as consumers of products whose origins are foggy at best, given the complex and difficult-to-trace global sourcing practices of many corporations. As a result, we know far too little about the conditions under which most products and services are generated—and what their social and environmental impacts are.

While digital communication allows trivial pictures or videos to go "viral" and command fleeting attention from millions of people, the information we need to make educated decisions is much harder to come by. Consumers interested in grabbing the latest must-have gadget or fashion item at a discounted price may have no way of knowing whether workers were injured or local ecosystems harmed in the process of manufacturing these items.

With a few welcome exceptions, such as the expansion of renewable energy or reductions in chronic hunger, many of the trends chronicled in this edition of *Vital Signs* are headed in a worrisome direction. Plotting these rising trends on a piece of paper or a computer monitor may give the illusion that they can just continue to rise. But a reckoning may not be too far off in the future, given resource depletion, pollution, and climate change.

Untrammeled consumerism lies at the heart of many of these challenges. As various articles in this edition of *Vital Signs* show, consumption choices matter greatly. Apart from environmental consequences, the health and livelihoods of many millions of people depend on the way in which diverse products like fish, meat, coffee, cotton, paper, or plastic are produced and traded.

Consumers often do not know the full footprint of the products they are buying, such as the embedded water in a T-shirt or a steak, the pesticide exposure of cotton farmers, the repercussions of unpredictable coffee prices for small growers, or the local devastation caused by timber companies cutting down forests to produce paper. Labeling and certification programs, although helpful, still only cover a limited range and portion of products.

Growing fish consumption has depleted many of the world's fisheries—so much so that fish catch has stagnated for the past 20 years. Meanwhile, fish farming is growing by leaps and bounds. Yet its massive expansion has triggered concerns about land and marine habitat degradation, pollution, and the spread of diseases among fish populations raised under the crowded conditions of intensive fish farming.

Global meat production has more than quadrupled in the last half-century. People in industrial countries continue to eat more than twice as much meat as people elsewhere. And because beef is far more water-intensive than other types of meat, the consumption habits of people in Argentina, Brazil, New Zealand, Australia, and the United States have a particularly heavy impact. The steady growth of global meat production and use has considerable environmental and health costs, given its large-scale draw on water, feedgrains, antibiotics, and grazing land.

World coffee production and consumption have doubled since the late 1970s, but there are concerns about working conditions as well as about the use of agrochemicals, deforestation, and impacts on biodiversity. Multiple initiatives seek to promote fairer trade, worker welfare, and organic growing methods. Coffees certified as sustainable represent no more than about one-tenth of global coffee trade, but at least they are growing rapidly.

The legions of small cotton farmers in the world face challenges largely beyond their control—among them are unfair subsidies by wealthy countries and volatile prices. Cotton is a very pesticide-intensive crop; pest resistance, adverse health impacts, and water pollution are common repercussions. Cotton's water footprint can be considerable—producing a pair of jeans takes as much as 10,850 liters of water. As is the case with coffee, several initiatives seek to push social and environmental standards, but they still account for a small share of cotton production.

Too often, paper products are discarded soon after their purchase, and only a portion is recovered for recycling, even though that would save trees, energy, and water. More than half of all paper produced is used for wrapping and packaging purposes. "Junk mail" is all too ubiquitous. Although paper consumption is shifting toward Asia, people in wealthy western countries still use far more on a per capita basis than everybody else.

Plastics are found everywhere—from transportation and construction to health care, the food and beverage sector, and consumer goods. Per capita use remains about five times larger in Western Europe and North America than in Asia. As with paper, unnecessary packaging accounts for a huge share. Recovery and recycling of plastics remain minimal in most countries, and millions of tons end up in landfills and oceans each year.

For these and other materials, it is essential to reduce short-lived and unnecessary usage and to find more environmentally friendly alternatives. Raising consumer awareness can help, but many changes need to happen long before products find their way onto store shelves. That requires action by governments.

But we live in a world where deregulation and privatization are the rule. A number of proposed new international trade and investment agreements will make it much easier for corporations to sue governments when they perceive that regulations might impinge on their profits. The likely result would be fewer social, health, or environmental rules and, presumably, less citizen recourse to the information consumers need about the impacts of products. Passing these agreements would further sacrifice sustainability on the altar of profits.

Energy Trends

Danny Rimpl

A lignite-fired power plant in Timişoara, Romania

For additional energy trends, go to vitalsigns.worldwatch.org.

Global Coal Consumption Keeps Rising, But Growth Is Slowing

Christoph von Friedeburg

Global coal consumption keeps rising, reaching 3,826.7 million tons of oil equivalent (mtoe) in 2013.[1] (See Figure 1.) This represents a 3-percent increase from the previous year, and it came on the heels of 2.6-percent growth in 2012.[2] But the pace of growth is down from 7.1 percent in 2010 and 5.4 percent in 2011, when economies rebounded from the Great Recession.[3] Consumption rose from 1,074 mtoe in 1950 to 2,261 mtoe in 1988, after which it leveled off at around 2,200–2,300 mtoe in the 1990s before resuming strong growth.[4]

Looking at recent developments by region, energy-hungry emerging economies have been driving the expansion in coal use since the beginning of this century. China used 1,933 mtoe in 2013, and India, 324 mtoe.[5] In contrast, coal consumption in the United States and the European Union (EU) is declining. These countries have been replacing part of their coal consumption with natural gas and renewable energy, although China is taking steps in the same direction. The United States used 455.7 mtoe in 2013.[6] The EU, at 285.4 mtoe, accounts for over 56 percent of the consumption of the Eurasian region; Russia uses 93.5 mtoe.[7] (See Figure 2.)

The International Energy Agency (IEA) projects world coal demand will reach 6,350 mtoe in 2040, but it expects the growth rate to drop to 0.5 percent annually, principally because of weaker demand in countries that belong to the Organisation for Economic Co-operation and Development.[8]

As a consequence of growing demand, worldwide coal exports have increased over the past decade, with the exception of a slight dip during the global downturn.[9] Australia's exports have risen steadily but have recently been overtaken by Indonesia, whose annual sales skyrocketed to more than 200 mtoe.[10] South Africa is effectively a swing producer between the Atlantic and Pacific coal markets; its main competitors for the European market are Russia, Colombia, and the United States.[11] (See Figure 3.)

Global coal prices, for example in northwest Europe, have been increasing since the 1980s—from $30–40 per ton to peaks of up to $148 per ton in 2008, although the prices fluctuate considerably.[12] They slumped during the downturn, and recovered to $121.50 per ton in 2011.[13] Only recently has an oversupply of coal put pressure on prices, dropping the price in 2013 to $81.20 per ton in Europe.[14]

The coal supply is getting "dirtier" as strong demand and lower prices create markets for coal with a lower energy content. In 2012, for instance, the average heat content of coal produced in the United States was about 23.4 megajoules per kilogram (MJ/kg); back in 2005, by comparison, it was 29.17 MJ/kg.[15] This means that more and more coal needs to be burned to generate the same amount of heat for a desired electricity output.

Christoph von Friedeburg is a research fellow at Worldwatch Institute.

The Asia-Pacific region accounted for more than 70 percent of global coal consumption in 2013.[16] Coal remains the dominant fuel there, accounting for more than 50 percent of the region's primary energy consumption.[17] Africa has the second highest share of coal, at just over 20 percent of total energy use.[18] Coal plays only an insignificant role in the Middle East and Latin America, where oil and natural gas and large-scale hydropower, respectively, are the main energy sources. And in North America and Eurasia, coal's share is below 20 percent.[19]

Since domestic coal production cannot meet demand in Asia-Oceania, coal imports in the region as a whole leapt from 45.2 mtoe to 446.4 mtoe from 1980 to 2012.[20] The biggest consumer is China. Coal demand there has almost tripled since 2000, rising from 683.5 mtoe to 1,933.1 mtoe in 2013—more than half of the global figure.[21] Coal's share in the country's energy mix stood at 67.5 percent in 2013.[22] To meet coal demand, the nation so far has been relying on its domestic production, increasing it by up to 10 percent a year.[23] But analysts doubt that this is sustainable for another decade or longer, as the mining sector is already grappling with infrastructure bottlenecks, and the coal deposits that are the easiest to mine have already been partly exploited.[24] Imported coal is becoming competitive. As a consequence, China's imports have outweighed its exports since 2009.[25] Australia accounted for 38 percent of the imports, Indonesia for 34 percent, and South Africa for 13 percent.[26] At more than 180 mtoe, imports account for around 8 percent of total consumption, and they are gaining a larger share.[27]

To diversify its energy sources, the Chinese government in December 2012 increased its solar energy target from 21 gigawatts (GW) to 40 GW of installed capacity by 2015, with at least 10 GW from distributed solar energy.[28] In fact, in terms of capacity, investments, and exports, China has become a new world leader

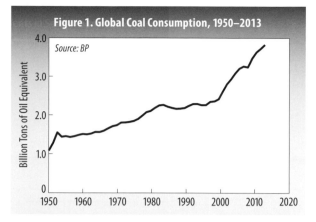

Figure 1. Global Coal Consumption, 1950–2013

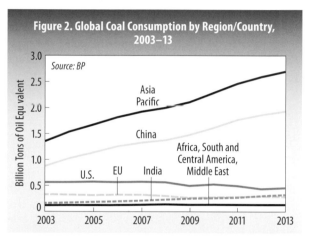

Figure 2. Global Coal Consumption by Region/Country, 2003–13

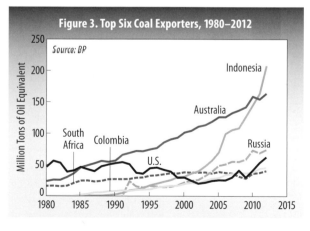

Figure 3. Top Six Coal Exporters, 1980–2012

in renewable energy technology.[29] Furthermore, China is looking into increased imports and domestic extraction of natural gas.[30] In addition, the government

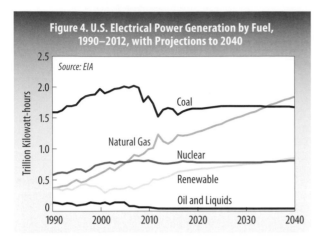

Figure 4. U.S. Electrical Power Generation by Fuel, 1990–2012, with Projections to 2040

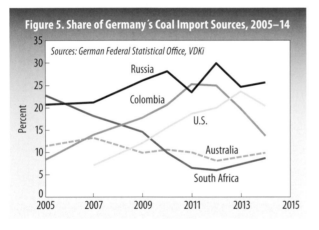

Figure 5. Share of Germany's Coal Import Sources, 2005–14

intends to reduce the nation's energy intensity—the energy consumption per unit of GDP—by 16 percent by 2015.[31]

In the United States, coal consumption has been in retreat since the start of the domestic shale gas boom. Coal's share in electricity generation declined to 39 percent in 2013, from 52 percent in 1990.[32] (See Figure 4.) In absolute figures, and despite the continued rise in energy demand, U.S. coal consumption decreased from 574.2 mtoe in 2005 to 455.7 mtoe in 2013.[33] During the same period, U.S. coal production decreased from 577.2 to 484.7 mtoe, with projections of a similar figure for 2014.[34] However, imports dropped drastically from 16.8 mtoe to less than 3 mtoe, while exports rose from 40.1 to 58 mtoe.[35] These trends could change in coming years if, as some analysts predict, many of the wells for hydraulic fracturing run dry and natural gas prices rise again or if substantial exports of liquefied natural gas begin.

Coal consumption in the EU has been on a marked downward trend since 1990. From a peak of 500 mtoe in 1987, demand fell to just above 300 mtoe from the late 1990s to 2008.[36] It dipped as low as 265 mtoe in 2009 before inching up to 293.4 mtoe in 2012 and then dropping back to 285.4 mtoe in 2013.[37]

There are two reasons for this trend. First, the EU's overall energy consumption has been nearly flat since 1990.[38] Energy intensity has shrunk to 69 percent of the 1990 value, while GDP is 47 percent higher than in 1990.[39] Second, coal's share in the EU primary energy consumption has shrunk from about 25 percent in 1990 to only 17 percent in 2013.[40] Contributing to the shift have been policies and financial incentives that raised the share provided by renewables to 14.4 percent in 2012, with a further target of 20 percent by 2020.[41]

Solid fuel imports into the 27 members of the EU in 2011 totaled about 144 mtoe, up from 116.1 mtoe in 1995.[42] Thus, dependence on imports has risen from 29.7 percent to 62.1 percent over the same period.[43] The explanation for the discrepancy between Europe's growing coal imports and its shrinking consumption is the phaseout of subsidies for coal mining, rendering European hard coal unprofitable. EU coal production has plunged from around 355.9 mtoe in 1990 and 270.9 mtoe in 1995 to 151.9 mtoe today.[44]

Recently, the U.S. share of EU coal imports has risen markedly, from below 14 percent in 2009 to around 18.3 percent in 2013.[45] But instead of adding to existing imports in Germany and other European countries, U.S. imports—and to some extent those from Russia—are replacing coal from other suppliers, such as Colombia and South Africa.[46] (See Figure 5.)

Factors like natural gas price changes, the profitability of using European lignite reserves, and midterm developments of the European Emission Trading Scheme could alter the trends currently seen.

The continued increase in coal consumption and the related carbon emissions cause substantial concern among climate scientists. If these trends continue, attempts to keep global warming below 2 degrees Celsius will likely fail.[47] Lacking meaningful, binding multilateral agreements on climate change, one source of hope is that the combination of decreasing energy intensity and declining costs of renewables will cause coal's share to keep shrinking as well as stop the global rise in the use of the dirtiest energy source.

Wind Power Growth Still Surging Where Strongly Supported

Mark Konold and Xiangyu Wu

Double-digit growth continued in the global wind market in 2013 as 35 gigawatts (GW) of additional capacity were added during the year, bringing the worldwide total to 318 GW—a 12.5-percent increase over 2012.[1] (See Figure 1). This was a significant drop from the average 21-percent annual growth rate over the last 10 years, however.[2] Overall investment declined slightly, from $80.9 billion in 2012 to $80.3 billion in 2013.[3]

Among the world's regions, the European Union is in the lead. Its 37-percent share of global installed capacity edges out Asia's 36 percent.[4] In fact, 16 European countries now have more than 1 GW of total installed wind capacity, compared with 4 countries in the Asia-Pacific region.[5] (See Figure 2). Once again, China installed more new capacity than any other country, with Germany, the United Kingdom, India, and Canada rounding out the top five countries in terms of added capacity in 2013.[6] (See Figure 3.) In 2013, China installed 16.1 GW of new wind power capacity, 24 percent more than it added the previous year.[7] By the end of 2013, total installed wind capacity there measured 91.4 GW, equivalent to 6 percent of the country's total electricity generating capacity.[8] The percentage of wind-generated electricity in China continued to rise in 2013, which shows that problems connecting wind farms to the grid are being resolved. (In 2011 alone, more than 10 billion kilowatt-hours [kWh] of wind power were lost because the grid lacked capacity to absorb it.)[9] In 2012, wind-generated electricity in China amounted to 100.4 billion kWh, or 2 percent of the country's total electricity output.[10] In 2013, those numbers rose to 134.9 billion kWh and 2.6 percent.[11]

Several European countries had significant additions in 2013. Germany added 3 GW to bring its total to 34.25 GW.[12] The United Kingdom installed nearly 2 GW of new capacity in 2013, much of which was offshore installations.[13] These two countries continue to be the region's most consistent leaders. France, Denmark, Italy, and Spain remain in the world's top group in terms of overall installed capacity, and they added 0.6, 0.6, 0.4, and 0.2 GW, respectively, in 2013.[14] Europe now has more than 117 GW of installed wind capacity, of which 34 percent is in offshore projects.[15] However, the European wind market slowed 8 percent in 2013 relative to 2012, and financing of new projects is becoming more challenging in response to policy uncertainty and declining incentives.[16]

In India, government policies in support of wind power have lapsed, and the country still struggles to eclipse the 3 GW mark set in 2011 for additions.[17] Only 1.7 GW were installed there in 2013.[18] To turn this tide, the Indian government reintroduced its generation-based incentive for wind and solar power projects between 100 kilowatts (kW) and 2 megawatts (MW), with an eye toward more

Mark Konold is Caribbean program manager in the Climate and Energy Program at Worldwatch Institute. **Xiangyu Wu** is a Climate and Energy intern.

robust growth in this sector in 2014.[19] Around $147 million will be put into the incentive in the 2013–14 budget, and low-interest loans for wind projects are also available through the government's National Clean Energy Fund.[20]

Other Asian countries are beginning to put more emphasis on wind energy development, though installations are not as eye-catching as in China and India. For example, Japan added 50 MW of new wind power capacity in 2013, leading to a total capacity of 2,661 MW, which is 0.5 percent of the country's total power supply.[21] Since the Fukushima accident in March 2011, Japan has focused on diversifying its energy supply. Offshore wind power, especially floating turbines, is promising for future development.[22]

South Korea's government has made "green growth" one of its national development priorities, but wind power still accounts for a small amount of the country's energy portfolio.[23] In 2013, some 79 MW of new onshore capacity was added, bringing the country's total to 561 MW.[24] However, the Korean government has targeted 2.5 GW by 2019 for offshore wind.[25]

The United States now has 61 GW of wind power capacity installed.[26] Policy uncertainty continued to affect the market as the Production Tax Credit (PTC), which needs to be renewed periodically by Congress, expired at the end of 2013. The United States experienced factory closures and layoffs due to the scarcity of new turbine orders. By the end of 2013, however, production capacity had increased significantly, with wind-related manufacturing taking place in 44 of 50 states.[27] The PTC was renewed in April 2014, so wind installation is expected to tick up in 2014 and 2015.[28] Also on the horizon may be comprehensive tax reform for renewable energy to lessen the "on again, off again" pattern that comes with annual incentive expiration.[29]

Sub-Saharan Africa, North Africa, and the Middle East saw only 90 MW of new wind power additions in 2013.[30] Taken together, these three regions have 1,255 MW of installed

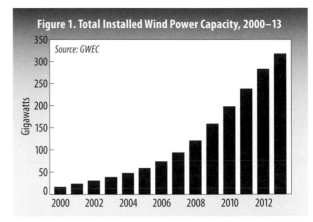

Figure 1. Total Installed Wind Power Capacity, 2000–13

Source: GWEC

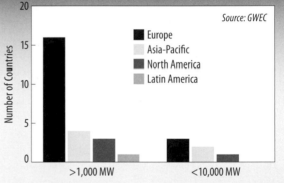

Figure 2. Installed Capacity by Number of Countries in a Region

Source: GWEC

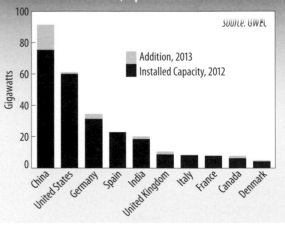

Figure 3. Cumulative Wind Power Capacity and Additions in 2013, Top 10 Countries

Source: GWEC

capacity.[31] Of this, 99 percent of the wind installations were found in just 9 countries: Egypt (550 MW), Morocco (291 MW), Ethiopia (171 MW), Tunisia (104 MW), Iran (91 MW), Cape Verde (24 MW), South Africa (10 MW), Israel (6.25 MW), and Kenya (5 MW).[32] However, the South African market was expected to expand in 2014, and several other countries have announced long-term plans to install commercial-scale wind power, including Ethiopia, Morocco, Kenya, Jordan, Tanzania, and Saudi Arabia.[33]

Continuing its drive to increase energy security and diversify supply, Latin America added almost 1.2 GW of new capacity, bringing the region up to 4.8 GW by the end of 2013.[34] Brazil alone added 953 MW of new capacity in 2013, and it contracted enough new capacity to have as much as 7 GW installed by the end of 2015.[35] A big factor in the region's wind expansion last year was the innovative approaches taken by Brazil and Uruguay. Both countries relied more on market-based solutions such as open tenders, contracts, and auctions instead of traditional government-backed programs based on federal funding or support from multilateral donors.[36]

Offshore wind continued to see impressive growth as projects became larger and moved into deeper waters. In fact, new machines in the 5–8 MW range are being tested for offshore use in Europe and Asia.[37] Until recently, deep-water offshore wind has developed on foundations adapted from the oil and gas industry, but deeper waters and harsher weather have become formidable challenges requiring newly designed equipment. British, Chinese, German, and South Korean shipbuilders are expanding to make larger vessels to transport bigger equipment and longer and larger subsea cables to more-distant offshore projects.[38]

These trends have kept prices high in recent years. As of early 2014, the levelized cost of energy (LCOE) for offshore wind power was up to nearly $240 per megawatt-hour (MWh).[39] This metric is a real-dollars gauge of technological competitiveness that includes the plant's full operational and financial life.[40] By comparison, the LCOE of onshore wind installations in various regions of the world is under $150 per MWh.[41] New floating designs are being tested to deal with these obstacles. Last year, Japan experimented with two 2-MW machines, hoping to quickly commercialize the technology.[42] In addition, the United Kingdom sought to expand its offshore momentum through leases for floating offshore wind solutions.[43] The aim is to drive down costs and revitalize old ports and related industries.

Back onshore, wind-generated power is becoming more cost-competitive against new coal- or gas-fired plants, even without incentives and support schemes. It is estimated that global levelized costs per MWh of onshore wind fell about 15 percent between 2009 and early 2014.[44] Over the past few years, capital costs of wind power have decreased because of large technological advances such as larger machines with increased power yield, higher hub height, longer blades, and greater nameplate capacity (which indicates the maximum output of a wind turbine—for example, a 2-MW turbine, under optimal conditions, produces 2 MW of power). In fact, the average size of turbines delivered to market in 2013 was 1.9 MW, up from 1.8 MW in 2012.[45] This upward trend will likely continue, as leading Chinese manufacturers, supported by government grants, are competing to develop turbines of 10 MW or larger.[46]

Tighter competition among manufacturers continues to drive down capital costs, and the positioning of the world's top manufacturers continues to shift. The top 10 turbine manufacturers captured nearly 70 percent of the global market in 2013, down from 77 percent the year before.[47] Vestas regained the top spot (see Figure 4), while GE Wind dropped to fifth largely because of a slower U.S. market. Goldwind rose to second place as Chinese manufacturers captured nearly one-fifth of the total global market.

In an effort to maintain profitability, manufacturers are trying new strategies, such as moving away from just manufacturing turbines. Some companies focus more on project operation and maintenance, which guarantees a steady business even during down seasons and can increase overall value in an increasingly competitive market.[48] Some manufacturers are also turning to outsourcing and flexible manufacturing, which can lower overall costs and protect firms from exchange rate changes, customs duties, and logistical issues associated with shipping large turbines and parts.

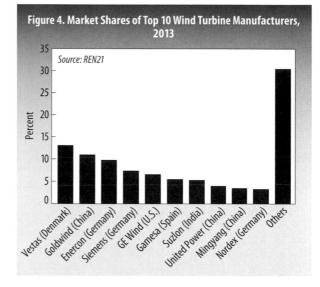

Figure 4. Market Shares of Top 10 Wind Turbine Manufacturers, 2013

Source: REN21

Solar Power Installations Jump to a New Annual Total

Max Lander and Xiangyu Wu

The year 2013 saw record-breaking growth for solar electricity generation as the photovoltaic (PV) and concentrated solar thermal power (CSP) markets continued to grow.[1] With over 39 gigawatts (GW) installed worldwide in 2013, the PV solar market represented one-third of all added renewable energy capacity.[2] Solar PV installations nearly matched those of hydropower and, for the first time, outpaced wind additions.[3] Global PV capacity topped 139 GW.[4] (See Figure 1.) Additionally, CSP grew by 900 megawatts (MW), bringing its capacity to 3.4 GW.[5] Solar PV and CSP together accounted for 18.7 percent of global net additions of power-generating capacity in 2013 from all types of energy.[6]

For the first time, Asia overtook Europe as the largest regional market, capturing more than 56 percent of the market share, while Europe came in second with around 29 percent.[7]

Despite the record growth in installations, global investments in solar electricity were down 20 percent, from $142.9 billion in 2012 to $113.7 billion in 2013, reflecting a significant decrease in costs.[8] (See Figure 2.) In July 2014, global PV module spot prices reached an all-time low of 63¢ per watt, a 10-percent decrease from the previous year.[9]

Photovoltaic continues to dwarf CSP capacity. Nine countries added more than 1 GW of solar PV in 2013, bringing the number of countries with more than 1 GW installed to 17, as Canada and Romania join the club.[10]

Adding 22.7 GW of PV, Asia dominated market activity for 2013.[11] China and Japan were the leading countries, followed at a distance by India (1.1 GW of installations), South Korea (0.4 GW), and Thailand (0.3 GW).[12]

China alone installed 12.9 GW of PV, the most ever installed in one year by any country.[13] The country's momentous expansion was fueled largely by its feed-in tariff (FIT) program, which supports large, grid-connected, utility-scale projects as well as distributed generation projects. By the end of 2013, Chinese PV capacity topped 20 GW, with more than 500 MW in off-grid installations.[14] However, the rapid pace of China's PV deployment created grid connectivity challenges.[15] China's share of global solar consumption trailed its share of installed capacity by 3.6 percent.[16] The government plans to develop 35 GW of PV capacity by 2015, with 20 GW of that total from distributed PV generation, marking a transition away from predominately utility-scale projects.[17] But with only 188 MW installed in the first quarter of 2014, China may not be able to reach this target.[18]

Japan came in second in global PV capacity additions with 6.9 GW, bringing its total capacity to 13.6 GW. As in China, this has been due to a strong FIT program that succeeded in attracting investment since its inception in July 2012.[19] Despite

Max Lander is a research assistant in the Climate and Energy Program at Worldwatch Institute. **Xiangyu Vu** is a Climate and Energy intern.

cutting solar FIT tariff rates in both 2013 and 2014, Japan's PV expansion has shown no signs of slowing, indicating a rapid drop in investment prices.[20] Australia's solar market installed its 1 millionth rooftop system in 2013, bringing the country's rooftop coverage to 14 percent.[21]

Europe installed close to 11 GW of PV and raised its total installed capacity to 81.5 GW by the end of 2013.[22] However, this represented the second annual decline in installations after peaking at 22.3 GW in 2011.[23] The region's share of global capacity additions plummeted to 29 percent in 2013 from 74 percent in 2011.[24] Germany led in new installations with 3.3 GW, followed by the United Kingdom, Italy, Romania, and Greece, each of which installed between 1 and 1.5 GW in 2013.[25] The most significant market declines occurred in Germany and Italy, while the remainder of the European market held constant with previous years. In Germany, a reduction of FIT rates and an increase in regulations for utility-scale projects contributed to the 56.5-percent fall in installations.[26] Italy's 61.5-percent decline in yearly installations was the result of replacing feed-in tariffs with additional tax rebates.[27]

North America added 5.2 GW of PV, bringing the continent's total to about 13.3 GW, or 9.5 percent of global capacity.[28] The United States installed the third most PV worldwide,

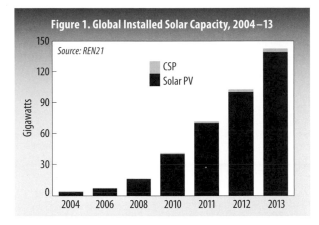

Figure 1. Global Installed Solar Capacity, 2004–13

Source: REN21

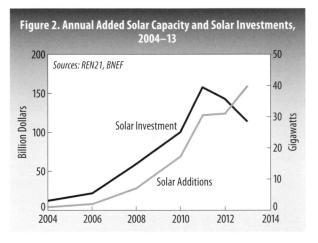

Figure 2. Annual Added Solar Capacity and Solar Investments, 2004–13

Sources: REN21, BNEF

with 4.8 GW, raising its total capacity to 12.1 GW.[29] California alone accounted for 55.1 percent, while Arizona, North Carolina, Massachusetts, and New Jersey represented another 36 percent.[30] Canada developed alongside the United States, with 444 MW of added PV, and Mexico added 45 MW.[31]

In Central and South America, solar development has been sluggish. Peru has installed the most PV in the region, and Chile and Brazil both have significant projects in the pipeline.[32] Despite consumption more than doubling in 2013, the region still accounts for a small fraction of the world's solar power.[33] The Middle East and Africa also had little PV activity, with the exception of Israel and South Africa, which added 420 MW and 75 MW, respectively.[34] South Africa added its first, and Africa's largest, PV power plant to the grid. PV markets in Jordan, Saudi Arabia, the United Arab Emirates, and Kuwait are expected to gain ground in coming years, as a number of tenders and purchase power agreements were signed in 2013.[35]

The CSP market also had another year of impressive growth, albeit from a still small base. The United States (adding 375 MW of capacity) overtook Spain (350

MW) for the first time since 2007.[36] The two countries together accounted for roughly 80 percent of the global market.[37] By the end of 2013, a total of 19 countries had CSP plants installed or under construction.[38] All other markets combined added 250 MW, tripling global capacity outside of the United States and Spain.[39] The principal additions were the 100-MW Shams 1 plant in the United Arab Emirates and the 50-MW plant in Rajasthan, India.[40]

Including both CSP and PV, solar power consumption increased by 30 percent globally in 2013 to 124.8 terawatt-hours (TWh).[41] Germany maintained its lead as the largest user.[42] As a region, Europe continued to capture the majority of global solar power consumption at 67 percent, followed by Asia with 23.9 percent and North America with 8.1 percent.[43] Worldwide, solar consumption equaled 0.5 percent of electricity generation from all sources.[44]

Nine European countries generated more than 1 TWh from PV: Germany, Italy, Spain, France, Greece, the United Kingdom, Czech Republic, Belgium, and Bulgaria.[45] (Spain also produced 4.6 TWh from CSP, remaining the only European country with significant CSP capacity.[46]) In the United States, total electricity generated by solar power more than doubled from 4.3 TWh in 2012 to 9.3 TWh.[47] (See Table 1.)

Table 1. Solar Capacity and Generation, Selected Countries				
Country	Solar Capacity by year-end 2013	Added Solar Capacity in 2013	Solar Generation	Share of Electricity Demand Generated by Solar
	(gigawatts)	(gigawatts)	(terawatt-hours)	(percent)
Germany*	35.7	3.3	29.7	5.3
Italy*	17.9	1.5	22.2	7.0
Spain	7.6	0.5	12.6	4.9
United States	13	5.2	9.3	0.2

*Photovoltaic only.
Note: Data for the United States and Spain include concentrated solar thermal power.
Source: REN21, Renewables 2014 Global Status Report (Paris: 2014); national reports from Germany, Italy, Spain, and the United States.

China continued to dominate PV manufacturing, with 64 percent of global production and 5 of the 10 largest manufacturing companies.[48] Collectively, Asia continued to have the lion's share of module production, with 86 percent.[49] (See Figure 3.) The share from Europe and the United States fell to 14 percent from 40 percent in 2007.[50] The biggest market mover was Japan, which increased its production by 24 percent due to the explosive growth of its domestic market.[51] While global PV module production increased modestly by only 3 percent over 2012, module shipments jumped by 24 percent, signaling an easing of oversupply problems.[52] Yingli Green Energy, Trina Solar, and Canadian Solar led the field in both production and shipments. (See Figure 4.) The share in global production of the top 10 manufacturing firms increased by 2 percent, climbing to 39 percent.[53]

Thin-film PV production continued to lose ground, falling to only 10 percent of global production (4.2 GW), compared with 19 percent in 2009, as it continues to be more expensive and less efficient than crystalline silicon technologies.[54] First Solar and Solar Frontier, an American and a Japanese company, accounted for 63 percent of all thin-film production.[55]

Moving forward, things are looking bright for solar development as prices continue to fall and approach grid parity in an increasing number of contexts. Rooftop solar is already less expensive per megawatt-hour than retail electricity in Australia, Brazil, Denmark, Italy, and Germany.[56] Estimates now also show that PV has become price-competitive without subsidies in 15 countries.[57] For 2014, solar installations were estimated to reach 40–51 GW.[58]

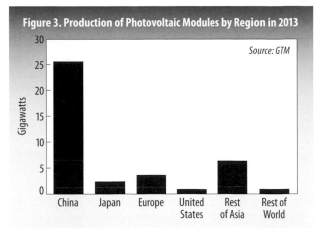

Figure 3. Production of Photovoltaic Modules by Region in 2013

Source: GTM

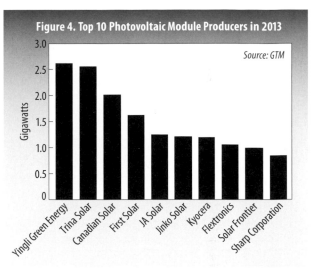

Figure 4. Top 10 Photovoltaic Module Producers in 2013

Source: GTM

Wind, Solar Generation Capacity Catching Up with Nuclear Power

Michael Renner

Advocates of nuclear energy have long been predicting a renaissance, yet this mode of producing electricity has been stalled for years.[1] Renewable energy, by contrast, continues to expand rapidly, even if it still has a long way to go to catch up with fossil fuel power plants, which account for roughly two-thirds of world electricity production.[2]

Nuclear's share of global power production declined steadily from a peak of 17.6 percent in 1996 to 10.8 percent in 2013.[3] Renewables increased their share from 18.7 percent in 2000 to 22.7 percent in 2012.[4] Hydropower was the leading source of renewable electricity (16.5 percent of global power in 2012), while wind contributed 3.4 percent and solar, 0.6 percent.[5] But wind and solar energy are the fastest growing electricity technologies worldwide. Between 2000 and 2012, wind power grew nearly 16-fold and solar jumped 49-fold.[6]

From its beginnings in the mid-1950s, global nuclear power generating capacity rose rapidly and reached 298 gigawatts (GW) in 1987, an average annual growth of 9.3 percent.[7] In the following 23 years, however, only 77 GW of capacity were added to reach 375.3 GW, at a rate of 3.4 percent per year.[8] From this 2010 peak, capacity declined to 371.8 GW in 2013, according to the International Atomic Energy Agency (IAEA).[9] Adverse economics, concern about reactor safety and proliferation, and the unresolved question of what to do with nuclear waste have put the brakes on the industry.

The IAEA's figures actually paint a rosy picture. A critical assessment by the World Nuclear Industry Status Report comes up with lower capacity figures and a more precipitous decline from a 2010 peak of 367 GW to just 333 GW as of July 2014.[10] The difference is due to assessments of the effective operational status of reactors, principally those in Japan after the Fukushima disaster; 38 of that country's 48 reactors have not produced electricity in two and a half years.[11] The Status Report uses a concept called long-term outage (LTO), under which "a nuclear power reactor is considered in LTO if it has not generated any power in the entire previous calendar year and in the first semester of the current calendar year."[12]

In stark contrast, wind and solar power generating capacities are now on the same soaring trajectory that nuclear power was on in the 1970s and 1980s.[13] (See Figure 1.) Wind capacity of 318 GW in 2013 is equivalent to nuclear capacity in 1990.[14] The 139 GW in solar photovoltaic (PV) capacity is still considerably smaller, but growing rapidly.[15]

Plotting nuclear, wind, and solar PV trends from a common starting point allows a closer comparison.[16] (See Figure 2.) The growth in wind capacity at first lagged behind the expansion of nuclear installations, but then it started to grow

Michael Renner is a senior researcher at Worldwatch Institute and codirector of *State of the World 2015.*

faster and is now outpacing nuclear. Solar PV capacity has entered a dramatic upswing even more quickly than either nuclear or wind did. The nuclear and wind industries each took 12 years to move from about 5 GW of cumulative capacity to about 100 GW; solar accomplished this feat in just seven years.[17]

A different picture emerges, however, with regard to electricity consumption from these sources. Although use of nuclear electricity has declined from a peak of 2,806 terawatt-hours (TWh) in 2006 to 2,489 TWh in 2013 (a slump of 11.3 percent), it still is four times larger than the use of wind power (628 TWh) and 20 times larger than solar power use (125 TWh).[18] (See Figure 3.) Yet consumption of electricity from all renewable sources—including hydropower, biomass, geothermal energy, and other sources—added up to 5,016 TWh in 2013, roughly double the amount of nuclear electricity.[19]

Wind and solar power are affected by a number of factors. There could be a time lag between completing the construction of a wind or solar farm and connecting it to the grid. In China, for example, solar PV capacity was added so quickly that it proved difficult to keep pace with grid connections, especially since many of the installations were in the sunny western parts of the country that are remote from the centers of electricity demand.[20]

A more fundamental factor is the intermittency of wind and solar resources. Compared with nuclear reactors, wind and solar facilities have a lower capacity factor—the amount of electricity actually produced compared with a facility's theoretical maximum. In the United States, nuclear reactors typically operate at between 85 and 90 percent of nominal capacity.[21] Solar PV plants tend to run at 16–28 percent and wind turbines at 30–45 percent.[22] However, renewable energy technologies continue to improve the effectiveness with which wind and sunshine are turned into power. The wind capacity factor has improved from 35 percent a decade ago to more than 50 percent for the latest designs.[23]

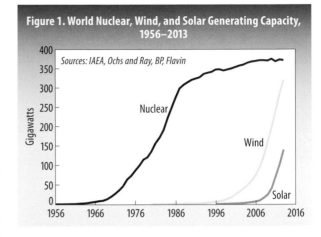

Figure 1. World Nuclear, Wind, and Solar Generating Capacity, 1956–2013

Sources: IAEA, Ochs and Ray, BP, Flavin

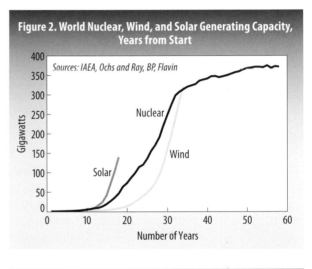

Figure 2. World Nuclear, Wind, and Solar Generating Capacity, Years from Start

Sources: IAEA, Ochs and Ray, BP, Flavin

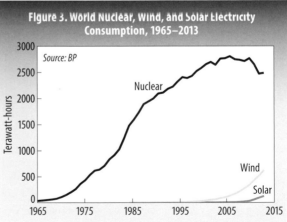

Figure 3. World Nuclear, Wind, and Solar Electricity Consumption, 1965–2013

Source: BP

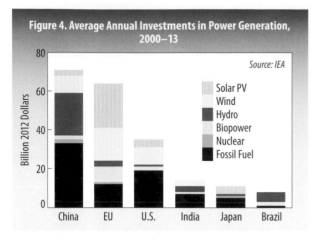

Figure 4. Average Annual Investments in Power Generation, 2000–13

Source: IEA

In recent years, renewable energy attracted far greater investments than nuclear power did. According to estimates by the International Energy Agency (IEA), nuclear investments averaged $8 billion per year between 2000 and 2013, compared with $37 billion for solar PV and $43 billion for wind.[24] Total renewable electricity investments averaged $153 billion annually during the same period, thus surpassing even the $106 billion spent on fossil fuel power.[25] Including $212 billion for transmission and distribution systems, an average of $479 billion was invested annually in the power sector as a whole.[26] Individual countries, of course, set diverging priorities, but nowhere did nuclear have a major role in power generation investments.[27] (See Figure 4.)

While renewable energy is poised to continue its expansion, the solar PV industry and, to a lesser degree, the wind industry have experienced considerable turbulence in recent years. Falling costs and regional shifts in manufacturing have forced a realignment and consolidation among manufacturers, although lower prices were welcome news for project developers and installers.[28]

Mostly due to lower costs for equipment such as solar panels and wind turbines, investments have actually fallen in recent years, although industry volatility and uncertainty about government policies also contributed to this development.[29] Total investments in wind energy grew from $14.5 billion in 2004 to a peak of $94.8 billion in 2010 but then declined to $80.3 billion in 2013.[30] Solar PV investments grew from $12.1 billion in 2004 to a peak of $157.8 billion in 2011, and now stand at $113.7 billion.[31]

In contrast with investment priorities, research budgets still favor nuclear technologies. Among members of the IEA (most European countries, the United States, Canada, Japan, South Korea, Australia, and New Zealand), nuclear power has received the lion's share of public energy R&D budgets during the last four decades.[32] (See Figure 5.) Nuclear energy (both fission and fusion) attracted $295 billion, or 51 percent, of total energy R&D spending between 1974 and 2012.[33] But this number has declined over time, from a high of 73.6 percent in 1974 to 26 percent today.[34] Renewable energy received a cumulative total of $59 billion during the same period (10.2 percent).[35] But in contrast to nuclear power's declining trajectory, spending on renewables rose from just 2.8 percent of energy R&D in 1974 to 20.8 percent in 2012.[36]

Worldwide, government R&D for renewables ran to an estimated $4.6 billion in 2013, with another $4.7 billion in corporate R&D.[37] The combined $9.3 billion is up from $5.1 billion in 2004, but it falls short of the $9.7 billion spent in 2011.[38]

Because they can be deployed at variable scales and constructed in less time, wind and solar power facilities are far more practical and affordable options for most countries than nuclear power reactors. Worldwide, 31 countries are operating

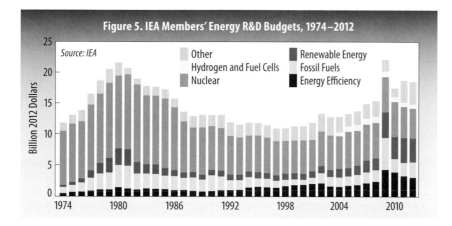

Figure 5. IEA Members' Energy R&D Budgets, 1974–2012

nuclear reactors on their territories.[39] This compares to at least 85 countries that have commercial wind turbine installations.[40]

A small number of countries rely heavily on nuclear energy. The United States, France, Russia, South Korea, and China together account for 68 percent of all nuclear electricity generated worldwide.[41] But several other countries have decided to phase out their reactors or to cease constructing new ones. Among the nations with nuclear reactors, six countries now generate more power from renewables (even when large hydropower dams are excluded) than from nuclear power: Brazil, China, Germany, India, Japan, and Spain.[42] In Spain, wind power—accounting for 21 percent of the country's total electricity generation—outpaced nuclear for the first time in 2013.[43]

The chances of a nuclear revival seem slim. Renewable energy, by contrast, appears to be on the right track. But it is clear that renewables have a long way to go before they can hope to supplant fossil fuels as the planet's principal electricity source. The expansion of sources like wind and solar will have to become even more rapid in order to stave off climate disaster, and that in turn means that their fate cannot be left to the whims of the market alone.

Smart Grid Investment Grows with Widespread Smart Meter Installations

Milena Gonzalez

Global investment in smart grids increased by $700 million from $14.2 billion in 2012 to $14.9 billion in 2013, and, for the first time, China outspent the United States.[1] A smart grid is a technology that improves the operation and reliability of electricity transmission and distribution systems by sending digital information in real time. The technology includes smart meters—which record energy use in short time increments and have two-way communication between the utility and its customers—and distribution automation, which allows remote monitoring and control of distribution systems to automatically fix faults and provide other adjustments.

China and the United States accounted for more than half of the global spending on smart grids in 2013. China invested $4.3 billion, most of which went into its smart meter program, while the United States invested $3.6 billion—33 percent less than in 2012.[2] (See Figure 1.) In the rest of the world, smart grid investment was modest in comparison, with growing investment expected in the near future in European and Asian markets.[3] While investment growth in smart grid technologies slowed somewhat in 2013 after very rapid increases in the past five years, growing levels of renewable energy integration and interest in higher grid reliability will continue to drive smart grid growth globally through 2020.[4] In fact, the global smart grid market is expected to cumulatively reach over $400 billion by 2020.[5]

By the end of 2013, China was on track to have 250 million smart meters installed, an 80-percent increase in just one year.[6] According to a 2010 census, there are 402 million households in China, so smart meters currently cover approximately 62 percent of households.[7] (See Figure 2.) China's metering program initially had the goal of installing smart meters in 95 percent of households by 2015, but the end

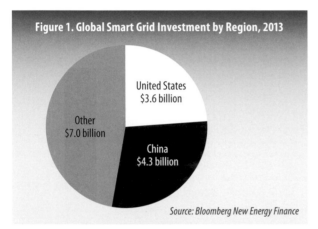

Figure 1. Global Smart Grid Investment by Region, 2013

United States $3.6 billion

Other $7.0 billion

China $4.3 billion

Source: Bloomberg New Energy Finance

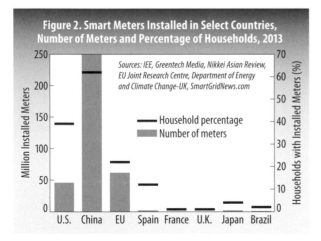

Figure 2. Smart Meters Installed in Select Countries, Number of Meters and Percentage of Households, 2013

Sources: IEE, Greentech Media, Nikkei Asian Review, EU Joint Research Centre, Department of Energy and Climate Change-UK, SmartGridNews.com

Household percentage
Number of meters

U.S. China EU Spain France U.K. Japan Brazil

Milena Gonzalez is a MAP Sustainable Energy Fellow with the Climate and Energy Program at Worldwatch Institute.

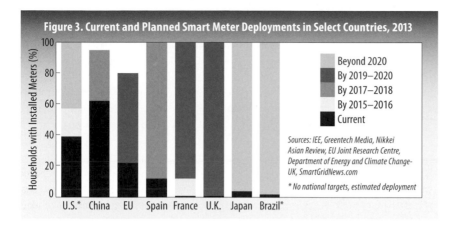

Figure 3. Current and Planned Smart Meter Deployments in Select Countries, 2013

date has been pushed back to 2017.[8] (See Figure 3.) While this may slightly slow down smart meter investment in China over the next few years, investment in other smart grid technologies, such as distribution automation, is expected to continue growing there.[9]

The U.S. decline in smart grid investment in 2013 came as the Smart Grid Investment Grant (SGIG) Program under the American Recovery and Reinvestment Act of 2009 wound down. As of October 2013, all of the $3.4 billion in federal funding provided under the act had been spent in cost-shared deployment of smart grid technologies and systems across the country.[10] This investment was matched with $4.5 billion in private funding, for a total investment of $7.9 billion since 2009 to support 99 projects.[11] The projects include electric transmission systems, electric distribution systems, advanced metering infrastructure, and customer systems. Notably, some of the federally funded projects included 65 smart meter projects that resulted in the installation of 14.2 million smart meters.[12]

Smart meters have also been deployed outside of the SGIG Program. As of July 2013, some 46 million smart meters had been installed in the United States, covering 40 percent of households in the country.[13] This was a 33-percent increase since May 2012.[14] (See Figure 4.) Besides modernizing the electric grid, strengthening cybersecurity, and improving interoperability, these and other smart grid projects have resulted in significant cost savings for utilities thanks to fewer and shorter power outages, higher energy and operational efficiency, and reductions of peak energy demand.[15] Furthermore, the collection of immense amounts of data from the smart grid is allowing utilities to optimize their operations and helping customers to better control their energy use and costs.

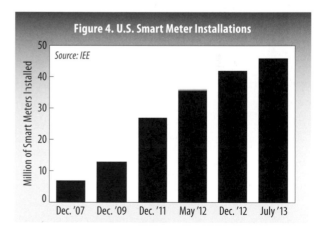

Figure 4. U.S. Smart Meter Installations

By the end of 2013, Europe had 459 smart grid projects across 30 countries, with the United Kingdom, Germany, France, and Italy leading in investments.[16] East European countries continued to lag behind in this field.[17]

In response to the Electricity Directive 2009/752/EC, which mandates European Union member states to install smart metering systems in 80 percent of households by 2020 where cost-effective, many countries have carried out cost-benefit analyses, most of which recommend deployment of the technology.[18] By the end of 2013, a total of 13 countries had developed smart meter roadmaps and 5 others were on track to do so as well.[19] Currently, approximately 22 percent of the European Union's 281 million metered customers have smart meters.[20] Until now, deployment has been led by Italy, Sweden, and Finland, but future investment is expected from Spain, France, and the United Kingdom. Germany, one of the larger potential markets, has rejected the Electricity Directive in favor of putting smart meters in selectively.[21] The cost-benefit analysis there showed that the installation costs for low-consumption households would be greater than the energy savings.[22]

Spain's national utility, Iberdrola, began using smart meters in pilot projects in 2010.[23] In 2012, it awarded a contract of $395 million to seven companies for the rollout of 1 million meters under its STAR project, which aims to upgrade 10 million meters by 2018.[24] Iberdrola's meter data collection system, which reads about 2 million meters, was completed in July 2013.[25] In France, the first phase of a nationwide smart meter rollout has begun.[26] By 2020, France aims to deploy 35 million smart meters, 3 million of which will be installed in the first phase by 2016.[27] Currently, there are about 300,000 smart meters installed in pilot projects.[28]

In the United Kingdom, the smart metering program aims to cover all homes and small businesses by 2020.[29] Energy suppliers are responsible for installing 53 million gas and electricity meters by then.[30] In September 2013, the British government awarded a license to establish and operate the smart metering Data and Communications Company, which will link each meter with energy suppliers, network operators, and energy service companies.[31] By December 2013 there were 296,000 smart meters installed in homes and 529,000 in non-residential sites in the United Kingdom, representing 0.6 and 18.7 percent of these sites, respectively.[32] The larger rollout of smart meter installations is expected to begin in late 2015.[33]

While far from the investment level that China has displayed, other countries in Asia are putting money into smart grid technologies and preparing for their use. South Korea, a global leader in this technology, is continuing to implement its smart grid roadmap and rollout of smart meters. In December 2013, in a submission to the United Nations Framework Convention on Climate Change, Japan proposed to install smart meters on every commercial, industrial, and residential building by the early 2020s as part of its strategy to cut carbon emissions.[34] This target builds on the contract that TEPCO, the largest utility in the country, was awarded last year for the deployment of smart meters to its 27 million customers by 2023.[35] Kepco, another Japanese utility, has already installed 2 million meters.[36]

In August 2013, India released a smart grid roadmap that includes the electrification of all households for a minimum of eight hours, demand response applications such as time-of-use tariffs for selected parts of the population, and reduction of grid losses by 2017.[37] The roadmap aims to launch smart grid pilot projects in 2015, to develop a low-cost, indigenous smart meter, and to target overall grid modernization.[38]

In Latin America, Brazil continues to lead the region in smart grid investment. In 2013, smart grid market revenue was $36 million, but it is expected to jump to $432 million by 2020 due to the need to upgrade the grid and improve reliability.[39] Currently, Brazil has over 1 million smart meters installed, but many more are expected from utilities in the near future.[40] In Barueri, for example, the city's utility plans to invest $32 million to install smart meters for its 60,000 customers by 2015.[41] Further, the government of Brazil has launched Inova Energia, a $1.4-billion plan to foster innovation in the electricity sector, which will support Brazilian companies and institutes in developing and commercializing technologies such as those used in smart grids.[42]

Smart grids are past the pilot phase and are increasingly being installed around the world. While China and the United States accounted for the lion's share of these investments in 2013, Europe, Asia, and Brazil will continue to drive growth. Higher levels of renewables connecting to the grid and the need for greater grid reliability will continue to push smart grid deployment.

Global Energy and Carbon Intensity Continue to Decline

Haibing Ma

Global energy intensity, defined as worldwide total energy consumption divided by gross world product, decreased 0.19 percent in 2013.[1] That may not seem all that impressive, but considering that energy intensity increased steeply between 2008 and 2010, this small decline continues a much-needed trend toward lower energy intensity, which basically means that people are using energy more efficiently.[2]

While a growing economy generally correlates with growing absolute energy use, energy intensity may well decline. During the 1970s, this is generally what happened. There were small bounces in 1975 and 1976, however, between the twin oil crises of 1973 and 1979.[3] (See Figure 1.) During an energy supply crisis, consuming nations tend to drastically restrict their energy consumption. But when the crisis eases, even if only temporarily, energy demand rises as efforts to boost the economy take hold.

In the 1990s, industrial economies started to turn to a new growth paradigm that relied heavily on service sectors. This "knowledge-based economy" is much less energy-intensive than the economic model that most nations adopted during industrialization. As a result, global energy intensity decreased 13.72 percent during the decade—the largest drop in the past 50 years.[4]

While industrial nations shifted their focus to developing service sectors, however, their heavy industries were partially transferred to emerging economies, like South Korea and China, causing the energy intensity in those countries to rise. Still, on a global scale, this trend was easily offset, allowing for a steady lowering of energy intensity throughout the 1990s.[5]

Things have turned out to be vastly different in the new millennium. The first decade saw great volatility, with two upward surges during 2002–04 and 2008–10, with average annual increases in those periods of 0.69 percent and 1.16 percent, respectively.[6] The period between 2004 and 2008, in contrast, saw a decrease in intensity of 3.50 percent.[7]

In the early years of that decade, large emerging economies like China started investing heavily in energy-intensive sectors, causing the global figure to increase. Then after the onset of the global financial crisis in 2008, many countries implemented massive stimulus packages, which usually focus on energy-intensive sectors like manufacturing, construction, and infrastructure. This largely accounts for the increase in global energy intensity between 2008 and 2010.[8] As the world economy began to recover after 2010, the previous pattern of global energy intensity reductions resumed. But because of these fluctuations, by 2013 energy intensity was only 1.82 percent below the level in 2000.[9]

As noted earlier, energy intensity is a ratio determined by both energy

Haibing Ma is a research associate in the Climate and Energy Program and the China program manager at Worldwatch Institute

consumption and gross economic product data. Gross world product is an aggregation of all nations' gross domestic product (GDP). To eliminate the fluctuations inherent in national currency exchange rates and inflation over time, it is better to use purchasing power parity (PPP) data based on real GDP values.[10] (See Figure 2.)

Using GDP values expressed in PPP still shows the same decreasing pattern overall, but the 15.87-percent drop between 2001 and 2013 is much more significant than the 0.94-percent decrease if PPP values are not used.[11] Another obvious difference is reduced volatility and a much smoother pattern. Two "bumps" are still visible, representing a slowing down or even a slight rise of global energy intensity during 2004–05 and 2009–10.[12]

Advanced economies such as the United States and Germany have followed the global trend of declining energy intensity.[13] By comparison, Japan's economy-wide energy efficiency shows minor but constant fluctuations between the mid-1990s and mid-2000s, a period when its domestic economy experienced difficulties.[14] (See Figure 3.)

The most turbulent trends were seen in newly industrialized and transitional countries. Russia, for instance, took a huge economic hit in the wake of political turbulence in the late 1980s and early 1990s, resulting in quickly rising energy intensity.[15] When its economy stabilized and resumed growing, energy intensity went on a rapid decline for most of the 2000s.[16] Energy intensity in South Korea had been growing for most of the 1990s as its national economy relied heavily on energy-intensive manufacturing sectors.[17] Then the country was hit hard during the Asian financial crisis in the late 1990s. Entering the new millennium, and especially since the mid-2000s, South Korea has focused more on green growth and sustainable develop-

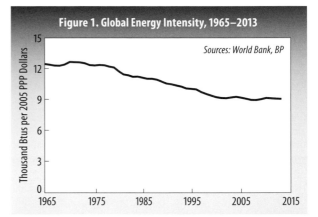

Figure 1. Global Energy Intensity, 1965–2013

Sources: World Bank, BP

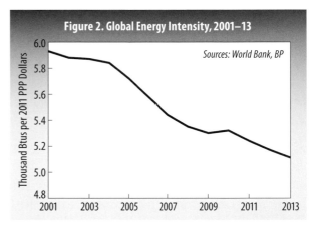

Figure 2. Global Energy Intensity, 2001–13

Sources: World Bank, BP

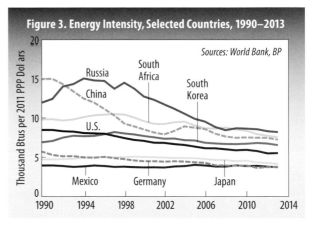

Figure 3. Energy Intensity, Selected Countries, 1990–2013

Sources: World Bank, BP

ment, which helped the nation achieve a steady decline in its energy intensity.[18]

China started from a very high level of energy intensity at the beginning of the 1990s.[19] As its economy grew at astonishing rates during the last decade of

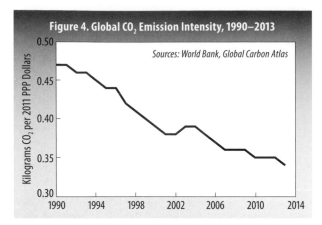

Figure 4. Global CO$_2$ Emission Intensity, 1990–2013

Sources: World Bank, Global Carbon Atlas

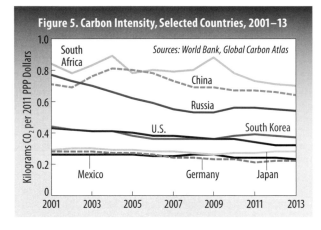

Figure 5. Carbon Intensity, Selected Countries, 2001–13

Sources: World Bank, Global Carbon Atlas

the twentieth century, China also achieved a sharp decline in energy intensity.[20] During that period, however, the economy eventually overheated and the nation also experienced some economic restructuring. As a result, China's economy-wide energy intensity increased between 2002 and 2004.[21] The government's enforcement of energy intensity reduction targets since 2006 has helped bring about a relatively steady decline in this indicator.

Carbon intensity is another important environmental indicator. It is defined as total emissions of carbon dioxide (CO_2) divided by gross world product. Global carbon intensity has followed the same general pattern of energy intensity.[22] (See Figure 4.) From 1990 to 2013, it dropped 36.62 percent.[23] But then it rose between 2002 and 2004.[24] After 2008, probably because of the impact of the economic recession, the decline in global carbon intensity generally slowed, but 2013 brought a slightly more rapid pace than in previous years.[25]

Advanced economies show a steadier declining trend in carbon intensity than newly industrialized and transitional countries do. (See Figure 5.) For instance, by the end of 2013 U.S. domestic carbon intensity was more than 60 percent below the 1990 level.[26] Germany currently has the lowest carbon intensity among industrial countries.[27] Japan, on the other hand, did not make much progress in lowering its emission intensity.[28] This may be due to the fact that the Japanese economy has been growing very slowly during the past two decades; in addition, Japan had achieved a quite advanced clean economy before the rest of the world, so there has been less room for improvement.

In 2006, China surpassed the United States as the world's largest CO_2 emitter.[29] Not coincidentally, China's carbon intensity also reached a temporary peak around 2004.[30] Given the need to act on climate goals, such an emissions-driven economic growth certainly cannot be sustained. Realizing the risk, the Chinese government has been taking aggressive efforts to slow its CO_2 emissions. In its climate action annual report released in November 2014, China claims its carbon intensity decreased 4.3 percent in 2013 and dropped 28.56 percent from the 2005 level.[31] World Bank data show lower drops of 3.61 percent in 2013 and 24.97 percent since 2005.[32] At a meeting in Beijing in November 2014, President Obama and President Xi issued a joint announcement in which China proposed to peak its carbon emissions by 2030.[33] The critical question is at what number this peak will be achieved. Further reducing the economy's carbon intensity will help to achieve a lower peak than otherwise possible.

Global energy intensity and carbon intensity are essentially measuring the efficiency with which human economic activities interact with nature. To ensure a sustainable development path globally, these two indicators need to be watched closely.

Effects and Sustainability of the U.S. Shale Gas Boom

Christoph von Friedeburg

Currently, only three countries are producing shale gas through hydraulic fracturing (fracking) on a commercial scale: the United States, Canada, and China. The United States is by far the dominant producer, with a new high of 32.9 billion cubic feet (bcf) per day in 2014.[1] Key U.S. regions for shale gas mining are Pennsylvania, Louisiana, and Texas. Canada is a distant second, with 3.9 bcf per day of natural gas as of May 2014; most of that production takes place in Alberta and Saskatchewan.[2] China is in third place, with currently only 0.25 bcf per day.[3] The Chinese shale gas fields are located in the Sichuan Basin. All three of these countries increased their output in 2014—and at a higher rate than conventional gas.[4]

The United States is the only country where shale gas production not only accounted for a significant share of total natural gas production but, by the end of 2013, surpassed daily output from non-shale wells and became the dominant source.[5] (See Figure 1.) This made the United States a leading nation in natural gas in spite of declining conventional production.[6] (See Figure 2.) Russia still has the strongest natural gas reserves (see Figure 3), followed by Iran and Qatar.[7] Global estimated technically recoverable resources of shale gas are 7,299 trillion cubic feet (tcf).[8] These appear to favor a different set of countries: China (1,115 tcf), Argentina (802 tcf), Algeria (707 tcf), Canada (573 tcf), the United States (567 tcf), Mexico (545 tcf), and Australia (437 tcf).[9] However, continued investigations can substantially reduce or increase current estimates, in particular of the fraction that can be economically retrieved. Shale resource exploration efforts are being undertaken in several countries, including Algeria, Argentina, Australia, Colombia, Mexico, and Russia.[10] But, for various reasons, none of these have commercial-scale production so far.[11] Important factors include the reserve's depth

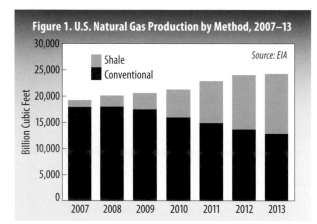

Figure 1. U.S. Natural Gas Production by Method, 2007–13

Source: EIA

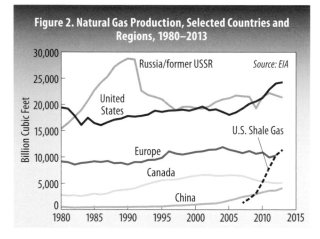

Figure 2. Natural Gas Production, Selected Countries and Regions, 1980–2013

Source: EIA

Christoph von Friedeburg is a research fellow at Worldwatch Institute.

below the surface, the surface material (rock or soil), existing infrastructure, and current natural gas prices in the markets.

Given limited domestic conventional reserves, most European countries depend on imports for natural gas; while some receive most of their gas from Norway or Qatar, many East European countries get half or more of their gas from Russia, shipped via pipelines through Ukraine.[12] The armed conflict in eastern Ukraine and European Union (EU) sanctions against Russia have led to calls for reducing or diversifying Europe's dependence on foreign energy sources.[13] But, in many cases, recoverable quantities of shale gas in Europe remain uncertain, and supplies are located deeper underground, partly in densely populated areas. Additional factors such as ownership of mineral rights, taxation, and substantial environmental and safety concerns inhibit development of Europe's shale gas resources, and in several countries there are ongoing discussions about fracking.

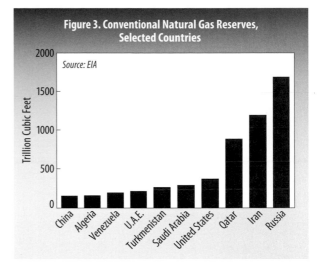

The British government appears to be in favor of exploiting this resource. However, just one shale well has been fracked so far, in Lancashire in 2011.[14] It caused two earth tremors, leading to a temporary ban that was in effect until 2012.[15] Since then, a few exploration wells have been drilled, but none have been actually fracked. A few companies are planning to apply for permission to proceed further.[16]

In Germany, technically recoverable shale gas reserves are estimated at 17.7 tcf.[17] In April 2015, the government submitted a draft bill allowing for exploratory fracking, with options for commercial fracking starting in 2019, provided favorable environmental and public safety assessments.[18]

Even in Eastern Europe, initial enthusiasm has given way to more sober assessments. The U.S. energy firm Chevron has ceased exploratory activities in Romania and Poland.[19] In Romania, contributing factors were reserves that are far lower and less profitable than expected, public protests about fracking, and lower oil prices rendering natural gas economically less viable.[20] In Poland, original estimates of more than 187 tcf were revised to just 12.4–14.7 tcf.[21] In Bulgaria, popular protests led the government to issue a ban on fracking.[22]

China's estimated reserves of 155 tcf of conventional natural gas are the largest in the Asia-Pacific region, but this is potentially dwarfed by the technically recoverable shale gas resources, which are about as much as those in the United States and Canada combined.[23] China has invested more than $1 billion in exploration so far.[24] But deposits are located in hard-to-access mountainous terrain at great depths. This makes the drilling, as well as establishing the needed infrastructure, such as roads and pipelines, more challenging and expensive. Another problem is the unavailability of the large quantities of water needed for hydraulic fracturing.[25]

Latin America has large recoverable shale gas reserves as well, but given problems with developing these resources, liquefied natural gas (LNG) imports increased 18 percent in 2014.[26] LNG imports into South and Central America totaled almost 700 tcf in 2013, mostly from Trinidad and Tobago.[27]

This leaves the United States and, to some extent, Canada as currently the only significant producers of shale gas. Projections have the United States turning its current net imports of natural gas into net exports by 2016, and optimistic forecasts foresee production overtaking consumption so much that exports could reach 18 percent of consumption by 2040, while consumption is steadily increasing.[28] Shale production has already lowered the market price of natural gas in the United States. It had increased steadily since the 1960s to a high of $12 per thousand cubic feet for commercial consumers in 2006, before the shale boom began.[29] Since then, it has dropped to $8–$9 per thousand cubic feet.[30] (See Figure 4.)

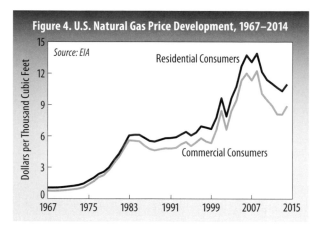

Figure 4. U.S. Natural Gas Price Development, 1967–2014

In the EU, natural gas import prices dipped to about $7.5 per thousand cubic feet after the recession, then climbed again to more than $13 in 2013.[31] Since then, a rise in coal consumption and a steady increase in the availability of renewable energy have crowded natural gas out of the electric power sector, reducing demand. By January 2015, the price had dropped to less than $8.5 per thousand cubic feet.[32] Various factors are influencing European prices and making predictions difficult, including the duration and severity of winters and uncertainty about the reliability of supplies from Russia.

In the United States, there was a substantial shift in fuels for electricity generation as natural gas became cheaper than coal. From as low as 16 percent in 2000, the share of natural gas rose to as high as 30 percent in 2012.[33] This happened at the expense of coal, whose share declined from 52 percent in 2000 to a low of 37 percent in 2012.[34] By 2040, natural gas is forecast to account for 35 percent of all fuel sources in power generation, topping coal at 32 percent.[35]

The marked shift from coal to gas in power generation helped reduce U.S. greenhouse gas emissions. But the global benefit of this reduction is dubious, because growing amounts of U.S. coal have found their way to export markets, notably to Europe. There, U.S. coal replaced supplies from other countries as well as domestic production, which is increasingly unprofitable.[36] But imported coal also became cheaper than natural gas. Coal's share in electricity generation, which had been in steady decline since the 1960s, started rising again in 2011, although 2014 data suggest that in the EU this short-term trend may have come to an end already.[37]

An intense debate has arisen about possible U.S. natural gas exports as well as their destinations. In most cases the gas needs to be liquefied in large facilities, transferred to shipboard tanks, and re-gasified at the destination. These processes add up to $4 per thousand cubic feet to the cost, making competition with locally available pipeline imports more challenging and reducing profitability.[38] The prospect of substantial LNG exports has raised concerns within the United

States that domestic natural gas prices could be driven up again, making energy more expensive.

Global consumption of natural gas (including LNG) reached about 118 tcf in 2013 and may have surpassed 120 tcf in 2014.[39] Most regions are increasing their consumption, although Europe's record is unsteady. (See Figure 5.) Today, only one U.S. LNG export terminal is in operation to help meet overseas demands.[40] Of 14 proposed terminals, 4 have already received permits.[41] These will have a combined capacity of about 3 tcf per year (8.3 bcf per day) and are expected to start operating between 2015 and 2019.[42] Australia´s projects under construction could supply about 2.8 tcf per year.[43] This compares with demand of about 15.4 tcf in the EU and 5.8 tcf in China in 2013.[44] At the moment, China's gas production covers roughly two-thirds of demand, with reserves allowing for significant production increase.[45] In addition, a 2014 long-term deal with Russia is likely to secure a substantial supply, so China does have a range of options beyond LNG imports from the United States.[46]

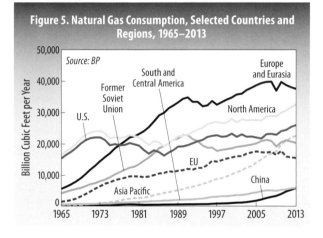

Figure 5. Natural Gas Consumption, Selected Countries and Regions, 1965–2013

A key question in each country is the impact of shale gas mining on local economies and environments. In the United States, shale gas development has been aggressively touted as a boon for jobs.[47] But logging and mining combined employ only 0.6 percent of Pennsylvania's workforce, for example, and employment there barely managed to get back to pre-recession levels—just as it did in some non-shale states.[48] Moreover, most jobs are temporary ones during the construction of the well pad and the drilling of the well; once the well is fracked and the gas is flowing, only a few employees are needed for maintenance.[49] In addition, 40 percent of the jobs went to skilled workers from all over the country, who live in trailers and motels and whose salaries only temporarily benefit local retail stores and restaurants.[50]

Studies show that development of clean energy projects can create many more jobs, and more-sustainable jobs, than shale development.[51] Moreover, damages to local roads from the fleets of trucks needed for well construction and wastewater transport amount to hundreds of millions of dollars.[52] Emissions of air pollutants from trucks and well pad diesel generators harm human health. And the often careless treatment of the toxic wastewater sludge caused by various additions to the fracking fluid pumped underground presents a long-term danger to the soil and aquifers. Finally, contrary to expectations, the U.S. shift of electricity generation from coal to cheap domestic natural gas has not led to a decrease in electricity prices that might have benefited the economy as a whole. U.S. average residential and commercial rates have continued their steady growth—from 10.65¢ and 9.65¢ per kilowatt-hour, respectively, in 2007 to 12.5¢ and 10.75¢ per kilowatt-hour in 2014.[53]

Furthermore, the future of U.S. shale gas production is unclear. Not only could initial reserve estimates be too high, but current predictions of actual production from the reserves could turn out to be too optimistic.[54] Recent studies from other sources, working with different parameter values for the various assumptions underlying any shale gas projection, are forecasting a peak in production in 2020, or even as early as 2016, and a tailing off thereafter.[55] This could call into question any substantial gas exports five years from now and even the sustainability of the U.S. shift to gas for electricity generation. It also indicates that any country's strong reliance on open-ended shale gas bounties, domestic or foreign, is unwise.

Environment and Climate Trends

Len Radin

Retreating Knik Glacier east of Anchorage, Alaska

For additional environment and climate trends, go to vitalsigns.worldwatch.org.

Greenhouse Gas Increases Are Leading to a Faster Rate of Global Warming

Joel Stronberg

Carbon dioxide (CO_2) is the main contributor to climate change. In 2013, the global combustion of fossil fuels and the production of cement resulted in the emission of 36.1 billion tons of CO_2, which was 61 percent over 1990 levels.[1] (See Figure 1.) (In terms of carbon, the emissions totaled 9.8 billion tons in 2013.) According to the most recent estimates, emissions in 2014 were projected to be 2.5 percent over 2013 levels.[2] In physical terms, this translates into the release of 37 billion additional tons of CO_2 in the atmosphere.[3] It is estimated that to keep the rate of Earth's warming below the 2 degrees Celsius threshold believed to be the temperature increase that will have severe and irreversible global environmental effects, total future emissions cannot exceed 1,200 billion tons.[4]

According to the International Energy Agency (IEA), even "meeting the emission goals pledged by countries under the United Nations Framework Convention on Climate Change (UNFCCC) would still leave the world 13.7 billion tons of CO_2—or 60 percent—above the level needed to remain on track for just 2°C warming by 2035."[5] The IEA's estimate will likely prove conservative, as most countries are, in fact, not on track to meet their stated reduction targets.[6] The 2 degrees Celsius threshold could be crossed before 2041, absent significantly more aggressive global mitigation actions. At the current rate of atmospheric emissions of CO_2 equivalents, the "quota" of 1,200 billion tons would be used up in less than a generation, although there is some disagreement on the exact timing of when this will happen.

The recently released *Emissions Gap Report* by the U.N. Environment Programme is a further warning: "In order to limit global temperature rise to 2°C and head off the worst impacts of climate change, global carbon neutrality should be attained by mid-to-late century."[7] Reaching carbon net neutrality between 2055 and 2070 is, according to the report, essential to minimizing the risk of severe, pervasive, and in some cases irreversible climate change impacts.[8]

The recent U.N. report and those by most of the scientific community reach four central conclusions: Time is of the essence; fully exploiting emission reducing actions now will prevent the need for significantly more severe and costly efforts in the future; the technological wherewithal to meet or exceed greenhouse gas (GHG) reduction targets exists; and linking development activities with proactive climate change actions will produce positive environmental and economic benefits for all.

As in 2013, the primary emitters in 2014 from the combustion of fossil fuels were expected to be China (28 percent), the United States (14 percent), the European Union (EU) (10 percent), and India (7 percent).[9] The picture in terms of

Joel Stronberg has been a consultant in the sustainable energy and environmental field for over 30 years.

emissions per person looks different, however, with the United States ranked first, with more than twice the per person emissions as China, ranked number two.[10] (See Figure 2.) There is a continued geographic shift in emissions from industrial to developing countries. For the first time, per capita emissions in China—at 7.2 tons per person (t/p)—exceeded the figure in the EU (6.8 t/p).[11] It is also estimated that despite India's currently low 1.9 t/p emissions, these will exceed those in the EU within five years.[12]

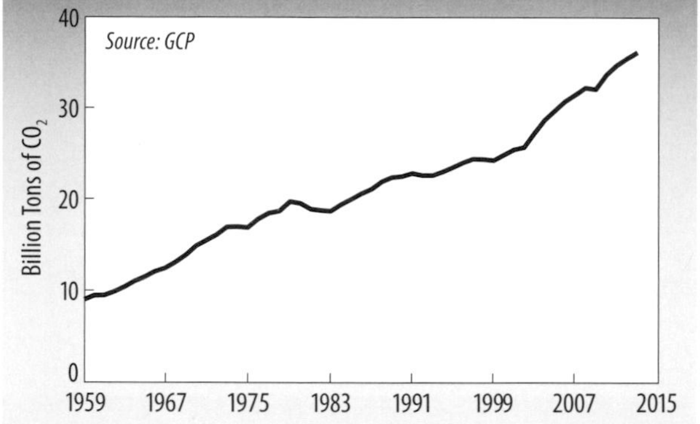

Carbon dioxide is the most prevalent GHG contributing to climate change, but it is not the only one. The three other major gases responsible are methane, nitrous oxide, and chlorofluorocarbons. The use of fossil fuels, particularly coal and petroleum, is the major source of CO_2, while natural gas production and agriculture are major contributors of methane. Methane is a super-potent GHG, trapping 86 times the heat of CO_2.[13] Satellite photos show that methane leakage from drilling and the pipeline delivery of natural gas offsets the CO_2 benefits attributed to using this energy source rather than coal.[14]

GHG emissions come from a variety of sources. Energy supply contributes 26 percent of emissions, industrial processes 19 percent, forestry 17 percent, agriculture 14 percent, and transportation 13 percent, while the building sector accounts for 8 percent and the waste and wastewater sector for the remaining 3 percent.[15] Flattening the GHG emissions curve so as to slow the rate of global climate change therefore requires a range of actions, including increasing the efficiency with which energy is produced, transmitted, and consumed; switching to renewable energy sources for electricity generation and transportation; and using non-fossil-fuel-based feedstocks for chemical production.[16]

The failure to meet previously established climate goals has occurred for a variety of reasons. Among them are the falling prices of coal, natural gas, and petroleum due to changed global economic conditions and the increasing supplies of oil and natural gas available through hydraulic fracking. The continued support for unsustainable practices, including direct and indirect fossil fuel subsidies, delays the transition to sustainable energy and means the 2 degrees Celsius threshold will be crossed sooner.[17]

Another factor is the slowdown in global economic growth and a movement toward austerity measures, causing some industrial nations to limit or consider

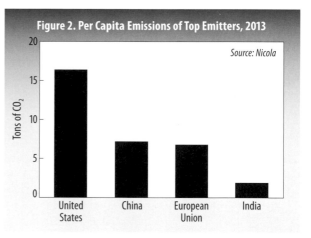

Figure 1. Global Carbon Dioxide Emissions from Fossil Fuel Combustion and Cement Production, 1959–2013

Source: GCP

Figure 2. Per Capita Emissions of Top Emitters, 2013

Source: Nicola

limiting near-term support for clean energy alternatives. This is particularly true in light of the currently depressed price of fossil fuels.

An expanding world population and the growth of developing-country economies contribute to the rising slope of GHG emissions. The continued growth of economies like China, India, and South Africa contributes to GHG emissions as it raises the global demand for power, cement, vehicles, the products of other energy-intensive industries, and minerals extraction. Even if the growth rates in countries are slowing, these nations are still adding environmental stresses when their growth is supported by the greater use of fossil fuels. And a growing world population increases food production needs. Industrialized farming and livestock practices were responsible for approximately 9 percent of total U.S. GHG emissions in 2012.[18] Methane and nitrous oxide are the two main pollutants released by the agricultural sector and are both super-potent GHGs.[19] Figure 3 illustrates the GHG emissions of CO_2 that are attributable to these practices in the United States, but the picture is similar in many countries around the world.[20]

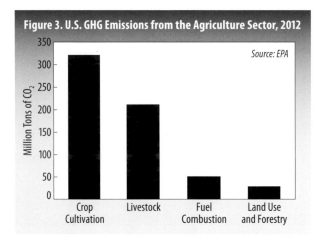

Figure 3. U.S. GHG Emissions from the Agriculture Sector, 2012

Source: EPA

In addition, deforestation not only generates carbon emissions from the burning of forest residues, it also reduces forests' capacity to capture carbon. It is estimated that halving the rate of deforestation and restoring significant acres of deforested lands worldwide by 2030 could reduce emissions by between 4.5 billion and 8.8 billion tons of CO_2.[21] The upper limit of the range would be equal to taking all the world's cars off the roads for a year.[22]

Increasing levels of GHG emissions may be an intractable problem in the near term. But the means to reduce emissions significantly and, consequently, the rate of climate change are available. The rapidly decreasing cost of clean energy alternatives such as solar and wind power reduces the need for government subsidies to make them competitive with fossil fuels.[23]

Innovative financing mechanisms are making solar systems more accessible to consumers everywhere. New storage technologies help address the problem of intermittency of wind and solar power. Particularly in remote areas of developing countries, mini- and micro-grids can be deployed quicker than building or expanding a centralized electric grid.

Some larger developing nations, including Brazil, China, India, Mexico, and South Africa, are increasingly interested in making sustainable energy industries a key element of their economies, both to meet internal energy demand and for technology exports. China's investment of $56.3 billion in renewable energy in 2013—mostly in solar and wind—made it the global leader for investment in utility-scale renewable projects.[24] In 2013, China also led the world in the manufacture of solar photovoltaic cells, accounting for nearly two-thirds of global production.[25]

Other actions that will slow rising GHG levels are found at both the corporate and the community level. Companies throughout the world are beginning to commit publicly to increasing their investments in decarbonization and reducing their carbon footprints. At the U.N. climate meeting in New York City in September 2014, for example, hundreds of billions of dollars of investment by 2020 were pledged to the United Nations Green Climate Fund by institutional investors, large insurance companies, and groups of major banks.[26]

Beyond establishing stringent aggregate emission targets at the December 2015 climate change conference in Paris, decision makers need information regarding the relative GHG emission intensities of various sectors such as agriculture, transportation, minerals extraction, and so on. Such data offer important insights on how best to meet national reduction targets consistent with each country's unique circumstances.

Increasingly the world has the means to achieve the needed emissions reductions. Whether or not the delegates to the 2015 climate conference can agree on a global accord, individuals, corporations, and national and subnational governments are showing greater willingness to undertake some needed actions. The important question is, Will the required steps be taken rapidly enough to avoid crossing the 2 degrees Celsius threshold?

Global Coastal Populations at Risk as Sea Level Continues to Rise

Vincent Yi

Global mean sea level has risen by 212.6 millimeters (mm) since 1880.[1] (See Figure 1.) And the rate of increase is accelerating.[2] (See Figure 2.) Overall, the global mean sea level has risen 1.65 mm per year since 1880.[3] But average sea level rise from 1993 to 2009 was almost double that long-term rate, at 3.2 mm per year.[4] This apparent acceleration is a matter of concern because some 10 percent of the world lives along a coast; as sea level continues to rise, these people will be threatened by further inundation by the sea and stronger storm surges.[5]

The evidence of climate change as a result of increases in greenhouse gas since the industrial revolution is readily evident, and the Intergovernmental Panel on Climate Change (IPCC) restated in its 2013 report that it is "extremely likely" these changes are driven by human actions.[6] The atmospheric carbon dioxide level, which did not rise above 300 parts per million (ppm) in the 650,000 years before 1950, currently sits at around 398 ppm.[7] This increase has led to stronger radiative warming of Earth, and the global mean surface temperature increased 0.85 degrees Celsius between 1880 and 2012.[8] Perhaps more alarming is the increase of 0.72 degrees Celsius just from 1951 to 2012.[9] This acceleration in surface temperature increase reflects acceleration in the rise in atmospheric carbon dioxide concentration. From 1980 to 2011, the concentration rose at 1.7 ppm per year, but the rate was 2 ppm from 2001 to 2011.[10]

The ocean absorbed over 90 percent of the excess heat energy from 1971 to 2010, and this increase in the ocean's energy content led to a rise in ocean temperatures.[11] In the top 700 meters of water, the ocean has warmed 0.302 degrees Celsius since 1969.[12] An increase in ocean temperature is critical because thermal expansion of ocean water is responsible for a substantial part of sea level rise.

Another main driver behind sea level rise is the melting of the global cryosphere—the planet's ice sheets, glaciers, and permafrost, as well as sea, lake, and river ice. Arctic sea ice extent decreased 3.5–4.1 percent a year from 1979 to 2012.[13] The decrease was even more dramatic for summer minimum sea ice extent, which decreased at a rate of 9.4–13.6 percent over the same time period.[14] Greenland ice sheets and glaciers around the globe have also been shrinking, and the rate of ice loss in both cases accelerated from the twentieth to the twenty-first century.[15] In contrast, Antarctica sea ice extent actually increased at a rate of 1.2–1.8 percent between 1979 and 2012.[16]

It is very likely that the rate of global mean sea level rise during this century will be higher than the rate observed between 1971 and 2010. In addition, the IPCC stated in its report that it is "virtually certain" that sea level rise will continue beyond 2100 due to continued thermal expansion in response to the warming ocean.[17]

Vincent Yi is an intern at Worldwatch Institute.

The two major factors influencing sea levels—ice loss and ocean temperature—are expected to increase throughout the twenty-first century under a business-as-usual scenario. By the end of the century, the top 100 meters of the ocean is expected to warm by 0.6 degrees Celsius under the most conservative scenario and 2 degrees Celsius under the business-as-usual scenario.[18] This is consistent with the IPCC conclusion that sea level rise will continue beyond 2100 due to thermal expansion.[19] The global cryosphere will also experience substantial ice loss during this time. Under a business-as-usual scenario, glaciers are expected to shrink by 35–85 percent by 2100, and the seas will be ice-free in September by mid-century.[20] Overall, global sea level is expected to rise 52–98 centimeters by the end of the century if humanity continues to use fossil fuels in the same way.[21]

The rise in sea levels will lead to permanent coastal inundation, increase the frequency and magnitude of floods along the coast, and affect drinking water and coastal agriculture through increased saltwater intrusion. Currently, more than half the people in the United States live in coastal watershed counties, and the population density in these counties is increasing.[22] Globally, around 10 percent of the world lives in a low-elevation coastal zone, which is defined as contiguous areas along a coast that are less than 10 meters above sea level, even though this area contains only 2 percent of the world's land.[23] (See Figure 3.)

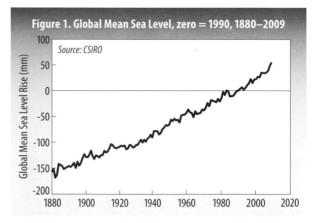

Figure 1. Global Mean Sea Level, zero = 1990, 1880–2009

Source: CSIRO

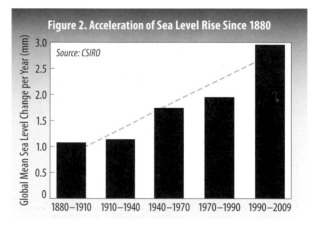

Figure 2. Acceleration of Sea Level Rise Since 1880

Source: CSIRO

Researchers estimate that an increase of just 0.38 meters in sea level would boost the number of people flooded by storm surges fivefold.[24] Industrial countries have demonstrated that they can recover fairly quickly from natural disasters, as Japan did after the 2011 earthquake and tsunami, but emerging economies do not have the same resilience.[25] This is especially of concern because some of the countries with the highest number and share of their population living in a low-elevation coastal zone are in developing Asian countries.[26] (See Table 1.) Sea level rise affects coasts differently, depending on local climatic and geographic conditions, but it is a threat faced by coastal populations worldwide. One developing Asian country highly susceptible to natural disasters is the Philippines. Over the past 30 years, the average sea level rise there was about 9 mm per year—triple the world average.[27] At one measuring station, sea level rose by 350 mm in the last 60 years.[28] When Typhoon Haiyan hit in 2013, the increased sea level amplified the storm surge and flooding, causing damages comparable to a tsunami.[29]

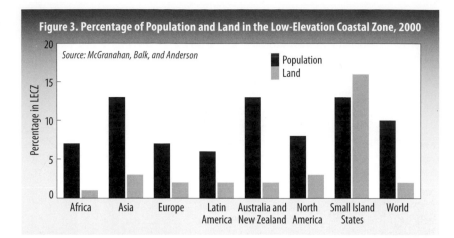

Figure 3. Percentage of Population and Land in the Low-Elevation Coastal Zone, 2000

Table 1. Total and Share of Population in Low-Elevation Coastal Zones in Asia

Country	Population	Percentage
	(million)	
Vietnam	43.05	55
Bangladesh	62.52	46
Japan	30.48	24
Philippines	13.33	18
China	143.88	11
India	63.19	6

Source: Gordon McGranahan, Deborah Balk, and Bridget Anderson, "The Rising Tide," Environment and Urbanization, April 2007.

Future sea level changes at the local level can be different due to land features and coastal circulation patterns, but any increase will likely lead to displacement and increased harm to the population. As a major coastal city with over 8 million residents, New York City will be one of the areas hit hard by sea level rise.[30] Fortunately, the city has a model plan for addressing climate change challenges: "A Stronger, More Resilient New York."[31] It calls for the fortification of coastal protections, which includes natural barriers such as wetlands and infrastructure such as levees. It also includes efforts to renovate infrastructure, to update building codes so builders use weather-resistant design and materials, and to improve the city's emergency response to threats associated with a rising sea level.

Along with adaptation to the expected rise in sea level, mitigation—that is, reducing greenhouse gas emissions—is necessary to avoid the dangers inherent in the business-as-usual scenario. In addition to reducing the extent of sea level rise, mitigation also slows the rate of increase, which will provide some time to implement adaptive measures. Delays to mitigation will only reduce options for a climate-change-resilient future, and extensive delays could lead to a scenario where adaptation limits have been exceeded.[32] Curtailing the use of fossil fuels and modifying land use to reduce greenhouse gas emissions associated with human activities must therefore be a part of a comprehensive plan to address sea level rise effectively.

Transportation Trends

Japanese bullet train at Fukuyama station

For additional transportation trends, go to vitalsigns.worldwatch.org.

Auto Production Sets New Record, Fleet Surpasses 1 Billion Mark

Michael Renner

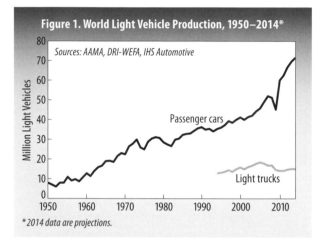

Figure 1. World Light Vehicle Production, 1950–2014*

Sources: AAMA, DRI-WEFA, IHS Automotive

Passenger cars

Light trucks

2014 data are projections.

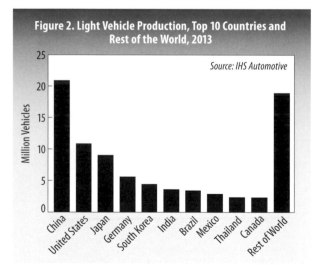

Figure 2. Light Vehicle Production, Top 10 Countries and Rest of the World, 2013

Source: IHS Automotive

Million Vehicles

China, United States, Japan, Germany, South Korea, India, Brazil, Mexico, Thailand, Canada, Rest of World

Michael Renner is a senior researcher at Worldwatch Institute and codirector of *State of the World 2015.*

Global production of automobiles keeps rising to new heights. London-based IHS Automotive put passenger-car production in 2013 at 69.6 million, up from 66.7 million in 2012, and projected a figure of 71.7 million for 2014.[1] (See Figure 1.) Adding light trucks (which in some countries, such as the United States, are used for passenger transportation), total light vehicle production rose from 81.5 million in 2012 to 84.7 million in 2013 and was projected to reach 86.8 million in 2014.[2]

IHS Automotive lists 51 countries as manufacturers of light vehicles.[3] But the leading five producers accounted for 60 percent of all light vehicles worldwide.[4] (See Figure 2.) China produced a stunning 20.9 million vehicles in 2013, up from 18.2 million in 2012.[5] The United States (10.9 million), Japan (9 million), Germany (5.6 million), and South Korea (4.5 million) follow at a considerable distance.[6] India (3.7 million), Brazil (3.5 million), Mexico (2.9 million), Thailand, and Canada (2.4 million each) complete the top 10 and contribute another 18 percent of the global total.[7] Seven other countries—Spain, Russia, France, the United Kingdom, Turkey, the Czech Republic, and Indonesia—produced at least 1 million vehicles each.[8]

The world's fleet of light-duty vehicles now surpasses 1 billion, having grown from 980.7 million in 2012 to 1,017 million in 2013— one light-duty vehicle per seven people.[9] Of that total, the passenger car fleet accounted for 739.8 million vehicles in 2013, up from 713.2 million in 2012.[10]

The United States has long been the world leader in motorization. The number of all motor vehicles per 1,000 people there rose from 324 in 1950 to a peak of 844 in 2007 but then decreased slightly to 812 in 2011.[11] Although automobiles are becoming more prevalent just about everywhere in the world, other countries and regions remain much less car-dependent, as an international comparison for 2011 shows.

Western Europe, for example, has almost the same number of vehicles (244.6 million in 2011 compared with the 248.9 million in the United States), but the vehicle-to-population ratio in Western Europe was equivalent to the ratio reached by the United States in 1972/73.[12] For Eastern Europe, the comparative year was 1951, but for many other countries it was even earlier.[13] (See Table 1.) If all countries had the same car density relative to population as the United States does, there would be 4.4 billion motor vehicles worldwide—more than four times the actual fleet.

Table 1. Motor Vehicles per Thousand People in 2011, and Comparison with the United States		
Country or Region	Vehicles per 1,000 People, 2011	Year When the United States Reached This Ratio
United States	812	2011
Canada	626	1974
Western Europe	590	1972/73
Eastern Europe	334	1951
Brazil	175	1925
Middle East	119	1922/23
China	70	1919
Africa	32	1916
India	20	1914/15

Source: Adapted from Stacy C. Davis, Susan W. Diegel, and Robert G. Boundy, Transportation Energy Data Book: Edition 32 (Oak Ridge, TN: Center for Transportation Analysis, Oak Ridge National Laboratory, 2013), pp. 3–7 to 3–10.

There are signs, however, that motorization in the United States may finally have peaked. The number of registered light vehicles reached a record 236.4 million in 2008 (when the economic crisis began).[14] The number of miles driven peaked in 2006, while fuel consumption topped out in 2004.[15] The fall in the amount of fuel consumed since 2004 is the result of a slightly smaller fleet, fewer miles driven, and higher fuel efficiency.[16] (See Table 2.)

Almost every tenth U.S. household—9.2 percent in 2012—does not have a vehicle, up from 8.9 percent in 2005.[17] In dense cities, the figure is much higher. In 2012, just over 56 percent of households in New York City did not own a vehicle, and in five other cities—Washington, DC; Boston; Philadelphia; San Francisco; and Baltimore—more than 30 percent of households were in the same situation.[18]

But many other U.S. cities lack the density, public transportation systems, walkability, and other factors necessary to make this a viable option. In 2011, some 86.3

Motorization Indicator	Peak Year	Value in Peak Year	Value in 2012	Decline since Peak
				(percent)
Light Vehicles (million)	2008	236.4	233.8	− 1.1
Miles Driven (billion)	2006	2773.0	2664.0	− 3.9
Gallons Consumed (billion)	2004	138.8	123.6	− 10.9
Light Vehicles per Person	2006	0.79	0.74	− 6.3
Miles Driven per Person	2004	9314.0	8488.0	− 8.9
Gallons Consumed per Person	2004	474.1	393.8	− 16.9

Table 2. Peak Years of Motorization in the United States

Source: Adapted and calculated from data in Michael Sivak, "Has Motorization in the U.S. Peaked? Part 5: Update through 2012," University of Michigan Transportation Research Institute, April 2014, Tables 1 and 3–5.

percent of all U.S. workers drove to their place of employment in private vehicles (and car-pooling represented only 10.2 percent of that total).[19] Public transportation, walking, and bicycling accounted for just 8.3 percent of these trips; all other means of transportation accounted for 1.2 percent, while 4.2 percent of employees worked from home.[20]

Vehicle fleets have either stopped growing or are growing very slowly in countries like Germany, France, Japan, and Canada.[21] In the 27 members of the European Union (EU), new-car registrations declined from a peak of 15.5 million in 2007 to 11.8 million in 2013.[22]

In many emerging economies, however, fleets continue to expand rapidly. The number of cars on China's roads skyrocketed from 3.8 million in 2000 to 43.2 million in 2011, and the country now has the third-largest fleet in the world, after the United States and Japan, having passed Germany's 42.9 million.[23] Russia's fleet grew from 20.4 million to 36.4 million, passing France and the United Kingdom (each with 31.4 million).[24] Brazil's fleet almost doubled, from 15.4 million to 27.4 million, during the same period of time.[25] India's nearly tripled, from 5.2 million to 14.2 million, passing South Korea's 14.1 million.[26]

Higher fuel efficiency is needed to limit automobiles' contribution to air pollution and greenhouse gas emissions. The current global average fuel consumption for all light-duty vehicles is 7.2 liters per 100 kilometers.[27] The Global Fuel Economy Initiative aims for a 50-percent improvement by 2050, but current trends fall short of achieving this goal.[28]

Better fuel economy means lower tailpipe carbon emissions. Japan, the EU, and India currently have the lowest allowable limits, at 119, 132, and 136 grams of carbon dioxide per kilometer driven (g CO_2/km), respectively.[29] Average emissions from new passenger cars sold in the EU decreased from 172g CO_2/km in 2000 to 127 grams in 2013.[30] Emission limits are still considerably higher elsewhere: 167–175 grams in South Korea, Brazil, Mexico, China, and the United

States. For light-duty vehicles in Australia, Mexico, and the United States, they are even higher, in the range of 199–203 grams.[31] (See Figure 3.)

Permissible emissions limits will be ratcheted down to 95 grams in the EU (by 2021); to 105 grams in Japan (by 2020); to 93 and 107 grams, respectively, for U.S. cars and light-duty vehicles (by 2025); to 113 grams in India (by 2021); and to a proposed 117 grams in China (by 2020).[32] Less ambitious goals are in place in Brazil (146.4 grams by 2017); South Korea (153 grams by 2015); and Mexico (153.1 grams for cars and 173.4 grams for light-duty vehicles by 2016).[33]

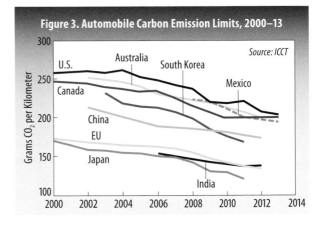

Figure 3. Automobile Carbon Emission Limits, 2000–13

One of the factors influencing fuel efficiency is the curb weight of vehicles. Cars in the United States are far heavier than those elsewhere—1,624 kilograms (kg) in 2011 compared with 1,393 kg in the EU; 1,291 kg in China; 1,160 kg in Japan; and 1,139 kg in India.[34] Within the EU, there is a wide range in vehicle weight, from 1,212 kg in the Netherlands to 1,484 kg in Germany and 1,562 kg in Sweden.[35]

In 1970, the amount of energy required for airplane travel in the United States was double that of driving—10,185 British thermal units (Btu) per passenger mile traveled versus 5,067 Btu.[36] But since then, trends in fuel efficiency and occupancy rates have favored air over car travel.[37] (See Table 3.)

Gasoline-powered cars predominate in the United States, China, and Japan. In contrast, diesel cars play important roles in India and Europe. In India, diesel vehicles claimed about half the market in 2012.[38] In Europe, diesels accounted for 55 percent of all new registrations in 2012, compared with gasoline cars at 42 percent.[39] As recently as 2000, in contrast, 69 percent of all vehicles sold in Europe were gasoline-powered and 31 percent were diesels.[40] All alternative-fuel vehicles

Table 3. Energy Consumption of Flying and Driving in the United States, 1970–2010			
Year	**Air Travel**	**Light Vehicles**	**Ratio**
	(Btu per passenger mile)		
1970	10,185	5,067	2
1980	6,029	4,628	1.3
1990	4,767	4,006	1.2
2000	3,892	3,926	1
2010	2,691	4,218	0.6

Source: Michael Sivak, "Making Driving Less Energy Intensive than Flying," University of Michigan Transportation Research Institute, January 2014, Table 1.

there combined accounted for just 2.4 percent in 2013.[41] Among them were 130,433 LPG vehicles; 81,574 natural gas cars; 31,105 plug-in hybrids; 24,173 electric vehicles; and 4,372 E85 (ethanol) vehicles.[42]

According to Navigant Research, sales of hybrid and electric vehicles worldwide topped 2 million in 2013.[43] The company forecasts that more than 6.5 million such vehicles will be sold in 2020.[44] In the United States, hybrids (495,000 cars) and electric vehicles (96,702) combined accounted for 3.8 percent of total sales in 2013.[45] In 2000, fewer than 10,000 hybrid vehicles were sold there.[46]

Toyota remains the leading hybrid manufacturer. The company's cumulative sales of hybrids, which began in August 1997, surpassed 6 million units at the end of 2013.[47] The main markets were Japan (2.8 million), North America (2.3 million), and Europe (647,000).[48] During 2013, the company sold nearly 1.3 million hybrids—10 times as many as in 2004.[49]

According to IHS Automotive, worldwide production of electric vehicles (battery electric and plug-in hybrids) expanded from 13,866 in 2010 to 242,075 in 2013.[50] The company forecast production of slightly more than 403,000 vehicles in 2014, up 67 percent from 2013.[51] The leading manufacturers are Nissan, General Motors, and Toyota, but Tesla's production is rising fast. The number of electric cars on the world's roads increased from nearly 100,000 at the beginning of 2012 to 405,000 units at the start of 2014.[52] Most of the cars are in the United States (174,000), Japan (68,000), and China (45,000).[53]

Alternative vehicles are slowly making inroads, but they are not yet significantly altering the resource and environmental impacts of automobiles. As electric vehicles become more numerous, a critical issue will be the source of the electricity that they run on—will it be generated from fossil fuels or from renewable energy?

Passenger and Freight Rail Trends Mixed, High-Speed Rail Growing

Michael Renner

According to the International Union of Railways (the UIC—from its name in French, the Union Internationale des Chemins de Fer), people traveled an estimated 2,865 billion passenger-kilometers (pkm) worldwide in 2013 by intercity rail.[1] The 2013 value is virtually unchanged from 2012 and confirms a slowing down since 2008.[2] From 1980 to 2008, passenger rail travel rose from 1,413 billion pkm to 2,687 billion pkm—3.2 percent per year—but from 2008 to 2013, the annual pace slowed to 1.3 percent.[3] (See Figure 1.)

Figure 1. World Passenger Rail Travel, 1980–2013

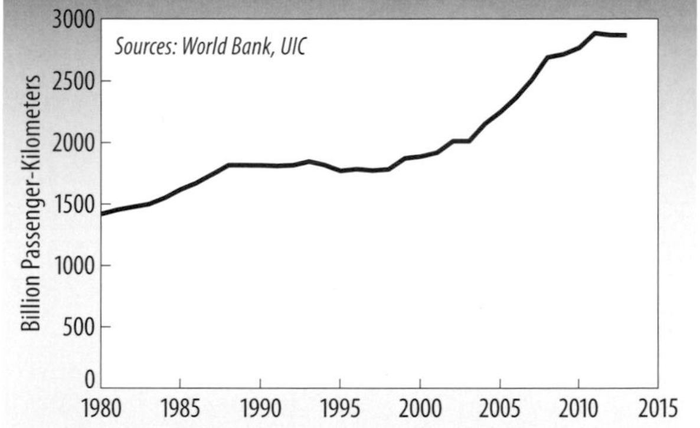

Sources: World Bank, UIC

Freight rail movements worldwide amounted to some 9,789 billion ton-kilometers (tkm) in 2013.[4] Freight rail expanded by 4.8 percent annually between 2000 and 2008.[5] Reflecting the impacts of the economic crisis, however, the 2013 figure is down about 4 percent from the peak value of 10,208 billion tkm reached in 2008.[6] (See Figure 2.)

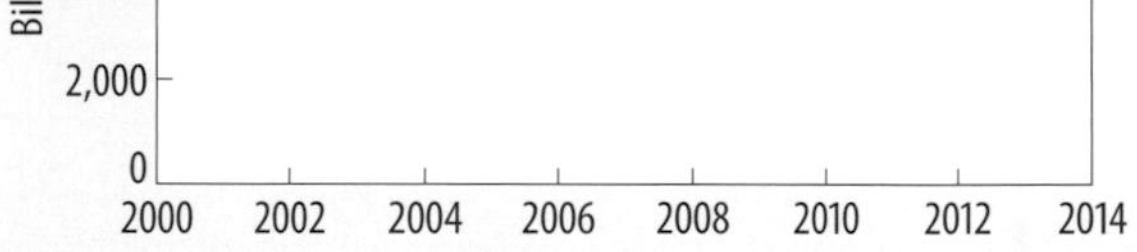

Even though more people and goods travel by rail, the length of the world's railway lines has not expanded nearly as much. UIC data indicate a total length for passenger and freight lines of 909,000 kilometers in 2000.[7] Growth during 2000–05 led the world's rail network to peak at 1.03 million kilometers.[8] Since then, however, the numbers have been flat.[9]

Figure 2. World Freight Rail Movements, 2000–13

Source: UIC

The world's rail vehicle stock runs to almost 3 million locomotives, railcars, and coaches—some 1.36 million in Asia (of which 750,000 are in China), 914,000 in Europe (with 125,000 in Germany and 121,000 in Russia), 667,000 in the Americas (with 442,000 in the United States), but just 46,000 in Africa.[10] However, the vehicle count is down from more than 3.6 million units in 2001.[11]

The capacity of freight rail vehicles has increased massively. In the United States, the number of freight cars declined from 1.4 million in 1970 to just 381,000 in 2012, yet the loads carried and especially the distances traveled increased, so

Michael Renner is a senior researcher at Worldwatch Institute and codirector of *State of the World 2015*.

that total ton-miles more than doubled from 1,231 billion to 2,756 billion during that time.[12] The amount of energy required to haul one ton over one kilometer dropped from 1,112 Btu to 473 Btu.[13]

Trains worldwide—whether they were transporting people or goods—traveled 12.7 billion kilometers (km) in 2013.[14] That is up from 10.3 billion km in 1991 and 9.7 billion km in 2001.[15] But individual trains have vastly different capacities. The number of railcars pulled by a locomotive may vary enormously. For passenger trains, seating capacity also varies considerably, especially since the introduction of double-decker cars.

Freight trains differ substantially in length and carrying capacity. For example, although the average U.S. freight train has 70 railcars, is almost 2 kilometers long, and hauls 3,000 tons of cargo, some freight trains there as well as in China, Brazil, or Australia, are made up of 300 railcars or more, pulling tens of thousands of tons of bulk cargo such as coal, iron ore, or grain.[16]

It takes a lot of people to operate and maintain rail systems, with nearly 7 million employees around the world in 2013.[17] The largest number of these, slightly more than 2 million, work in China, followed by India (1.3 million) and Russia (896,000); the members of the European Union together have just under 1.1 million employees, while other parts of Europe (excluding Russia) have another 523,000.[18] Asian countries, excluding China and India, employ 680,000 people.[19] Far fewer people work in the railways sectors of the Americas (273,000) and Africa (143,000).[20] Privatization and market liberalization policies over the last two decades brought enormous change, cutting total employment from 9.36 million in 1991.[21] In some countries, job loss has been dramatic. (See Table 1.)

Regionally, three-quarters of all passenger-kilometers are traveled in Asia and Oceania, up from less than two-thirds in 2000.[22] Europe's share has declined from 31 percent to 22 percent, while the Americas and Africa play minor roles.[23] India and China were the dominant countries in 2013, followed by Japan, Russia, France, and Germany.[24]

In freight rail, the regional picture is far more balanced. Asia and Oceania accounts for close to 37 percent, followed by the Americas (33 percent) and Europe (29 percent), while Africa has only 1.4 percent.[25] Since 2000, Asia's share of freight rail has gone up while that of the Americas has gone down.[26] In 2013, the United States, China, and Russia were far ahead of India, Canada, Ukraine, and Kazakhstan in total ton-kilometers.[27]

The longest rail lines are found in the Americas (37 percent in 2013) and Europe (34 percent).[28] Asia follows with 22 percent, and Africa with 7 percent.[29] The United States had by far the most extensive network in 2013, followed by Russia, China, India, and Canada.

Electrification of rail lines offers a number of advantages, including higher speeds, no need to carry fuel aboard, and higher energy efficiency.[30] Worldwide, 28 percent of rail lines are electrified, but the percentage varies enormously among individual countries—from as high as all the rail lines in Switzerland to as low as 2 percent in Indonesia.[31] (See Figure 3.)

The rail profiles of individual countries diverge considerably, depending on

Table 1. Rail Employment, Selected Countries and World, 1991, 2001, and 2013			
	1991	2001	2013
		(million jobs)	
China	1,998	1,453	2,032
India	1,651	1,545	1,328
Russia	1,877*	1,240	896
EU, of which:	1,143	703	1,082
Germany	442	182[†]	294
France	199	178	152
Poland	309	159	93
U.K.	139	22[†]	87
Italy	180	104	73
Spain	176	37	33
Ukraine	665	376	339
United States	206[†]	186	178
Japan	193	157	128
World	9,363	7,223	6,956

Soviet Union   † Partial data only (not including all rail companies)
Source: UIC, "Synopsis," Paris, 1991, 2001, and 2013.

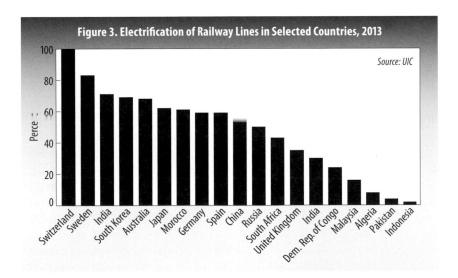

Figure 3. Electrification of Railway Lines in Selected Countries, 2013

Source: UIC

whether they are oriented mostly toward transporting people or goods. China, India, Germany, Russia, and Ukraine are among the top 10 countries worldwide in both categories.[32] (See Table 2.) But the United States, second in freight rail tonnage,

	Table 2. Top 10 Rail Countries, Passengers and Freight, 2013			
Country	**Passengers Carried**	**Rank**	**Tons of Freight Carried**	**Rank**
	(million)	(number)	(million)	(number)
Japan	8,819	1	31	-
India	7,651	2	922	4
Germany	2,008	3	390	7
United Kingdom	1,570	4	-	-
China	1,522	5	2,859	1
France	1,114	6	63	-
Russia	1,059	7	1,440	3
South Africa	531	8	197	-
Italy	513	9	39	-
Ukraine	485	10	457	6
United States	27	-	1,710	2
Brazil	n.a.	-	460	5
Canada	4	-	310	8
Kazakhstan	23	-	295	9
Australia	n.a.	-	242	10

Source: UIC, "Synopsis 2013," Paris, June 2014.

has only limited passenger rail traffic. Similarly, Japan—the undisputed leader in number of passengers transported—has small amounts of freight tonnage.[33]

High-speed trains now account for 12.5 percent of all passenger rail travel, up from 7.3 percent in 2004.[34] In Europe and East Asia, people traveled some 359 billion pkm on high-speed trains in 2013, more than double the 156 billion pkm a decade earlier.[35] (See Figure 4.) Japan and France were the top countries in high-speed rail until recently, together accounting for two-thirds of the world's high-speed travel in 2010.[36] But China has built the world's most extensive network of high-speed lines and grabbed the lead with a 40 percent share in 2013.[37]

Although different transportation modes are difficult to compare, available estimates indicate that rail transport is generally more fuel-efficient than movement by road vehicles. In the United States, intercity rail in 2012 required 2,481 Btu per passenger mile, compared with 3,193 Btu for passenger cars and 3,561 Btu for light trucks.[38] For freight modes, railroads required 13,800 Btu per freight-car mile in 2012, while heavy trucks used 21,525 Btu. However, the loads of railroad cars and trucks may vary and thus limit direct comparability.[39]

While rail is likely the more environmentally friendly choice in transportation in many situations, rail transport is not without its problems. Freight trains have long played a key role in moving raw materials from countries' interiors to export terminals, thus helping to maintain the resource-intensive global economy.

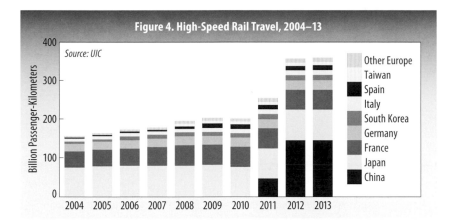

Figure 4. High-Speed Rail Travel, 2004–13

In the United States, growing amounts of oil are being moved by rail. According to Oil Change International, more than 800,000 barrels per day (bpd) of crude oil were transported in 2013, a 70-fold increase from 2005.[40] The capacity to load crude oil onto trains is already at 3.5 million bpd and could grow to 5.1 million bpd by 2016, leading to additional strong growth of oil shipments by rail.[41] The dangers are multiplying as well. The 117 spills of oil from train shipments in the United States in 2013 represented an almost 10-fold rise over 2008.[42] In Virginia, an April 2014 derailment spilled 30,000 gallons of crude oil into a river, and in Quebec, Canada, a derailment in July 2013 led to an explosion that killed 47 people.[43] Thus, while rail may be an environmentally friendly transport mode, the full impact depends very much on what is being hauled.

Food and Agriculture Trends

Neil Pa⌐er (CIAT)

Coffee plantation in Huila, in the Colombian Andes

For additional food and agriculture trends, go to vitalsigns.worldwatch.org.

Aquaculture Continues to Gain on Wild Fish Capture

Michael Renner

According to preliminary estimates by the United Nations Food and Agriculture Organization (FAO), total global fish production was expected to reach an all-time high of 160 million tons in 2013, up from 157.9 million tons in 2012.[1] This figure includes a projected wild capture of 90 million tons, down from 91.3 million tons in the previous year and from 93.7 million tons in 2011.[2] Wild capture has stagnated since the mid-1990s, reflecting the fact that many regional fisheries are at or even above their maximum sustainable catch levels.[3] In sharp contrast, aquaculture has expanded rapidly. Production from fish farming increased about 10-fold since 1984, reaching 66.6 million tons in 2012 and a projected 70 million tons in 2013.[4] (See Figure 1.) Aquaculture thus accounts for 44 percent of total fishery output (and for 49 percent of fish meant for direct human consumption), up from 8 percent in 1984.[5]

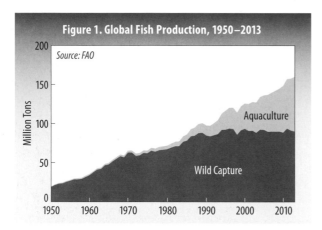

Figure 1. Global Fish Production, 1950–2013

Source: FAO

In the 1980s and 1990s, aquaculture grew by more than 10 percent annually, although the rate dropped to 6 percent in the decade after 2000.[6] A new World Bank report, *Fish to 2030*, predicts that 62 percent of food fish will come from aquaculture by 2030, with the fastest growth likely to come from tilapia, carp, and catfish.[7]

Global per capita consumption of fish from all sources increased from 18.9 kilograms (kg) in 2010 to 19.2 kg in 2012 (the latest year with detailed data).[8] Consumption of farmed fish reached an average of 9.4 kg in 2012.[9]

The majority of wild capture is from marine sources, accounting for 79.7 million tons in 2012, compared with 11.6 million tons for capture from inland waters.[10] The largest marine fisheries are the Northwest Pacific (21.5 million tons in 2012), the Western Central Pacific (12.1 million tons), the Southeast Pacific (8.3 million tons), and the Northeast Atlantic (8.1 million tons).[11] Together, these areas account for two-thirds of the world marine catch.[12]

Among marine producers, China is by far the leader, with 13.9 million tons in 2012, followed by Indonesia, the United States, Peru, Russia, Japan, and India.[13] (See Figure 2.) The top 10 countries account for 60 percent of the global total, and the next 15 countries haul in 23 percent.[14] China also tops the league of producers from inland waters, ahead of India and Myanmar.[15] (See Figure 3.) The

Michael Renner is a senior researcher at Worldwatch Institute and codirector of *State of the World 2015.*

top 10 countries had 70 percent of the world inland catch, and the next 15 countries took in 19 percent.[16]

Where China is truly dominant is in aquaculture production, controlling an astonishing 62 percent of the world total.[17] The top 10 producers account for 88 percent of the total; 7 of them are in Asia, with the remainder being Norway, Chile, and Egypt.[18] (See Figure 4.)

The bulk of farmed fish stems from inland aquaculture, which brought in 41.9 million tons in 2012, compared with 24.7 million tons from aquaculture in marine waters (known as mariculture).[19] Asia was responsible for 93 percent of world mariculture production.[20]

By live weight, 37 percent of the global total fish catch is exported.[21] In 2012, the value of exports of wild and farmed fish reached $129.1 billion, more than double that of a decade ago.[22] Preliminary data for 2013 suggest exports rose to $136 billion.[23]

Shrimp is the most-traded fish commodity (by value), accounting for 15 percent of the total value of international trade.[24] It is followed closely by salmon and trout (14 percent) and by groundfish such as hake, cod, haddock, Alaska pollock (each at 10 percent), and tuna (9 percent).[25]

The leading exporter in 2012 by far was China, followed at a distance by Norway, Thailand, Vietnam, the United States, and Chile.[26] (See Table 1.) These six countries accounted for just over 40 percent of global exports.[27] Data for 2013 are still incomplete, but they suggest that exports from China and Norway increased but those from Thailand decreased.[28] Imports present a different picture, with Japan and the United States the leading countries, followed by China, Spain, France, and Italy.[29] Preliminary data for 2013 indicate that the United States replaced Japan as the leading importer ($19 billion and $15.3 billion, respectively), while China (at $8 billion) remained in third place.[30] According to a World Bank projection, China's growing demand will account for 38 percent of worldwide fish consumption by humans by 2030.[31]

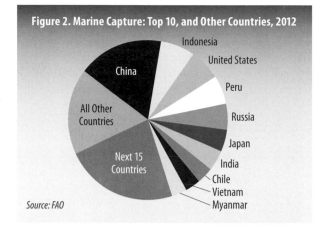

Figure 2. Marine Capture: Top 10, and Other Countries, 2012

Source: FAO

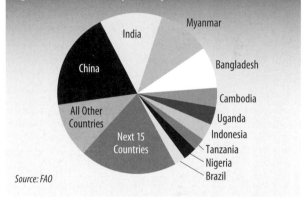

Figure 3. Inland Water Capture: Top 10 and Other Countries, 2012

Source: FAO

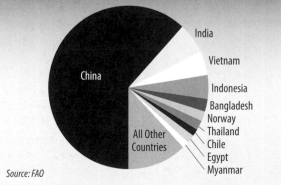

Figure 4. Aquaculture Capture: Top 10 and Other Countries, 2012

Source: FAO

Table 1. Leading Fish Exporters and Importers, 2012			
Exporters		**Importers**	
	(billion dollars)		(billion dollars)
China	18.2	Japan	18.0
Norway	8.9	United States	17.6
Thailand	8.1	China	7.4
Vietnam	6.3	Spain	6.3
United States	5.8	France	6.1
Chile	5.1	Italy	5.6

Source: FAO, "Fact Sheet: International Fish Trade and World Fisheries," February 2014.

In 2012, developing countries exported 61 percent of internationally traded fish by quantity and 54 percent by value.[32] FAO notes that net export revenues of $35.3 billion made this a larger source of income for developing countries than all other agricultural products combined, including rice, meat, milk, sugar, and bananas.[33]

Tens of millions of people engage in fishing operations worldwide. FAO puts the number of people working in the primary sector of capture fisheries and aquaculture at 54.8 million as of 2010, up from 30.9 million in 1990.[34] However, the share of these people employed in capture fisheries is decreasing (down to 70 percent in 2010 from 87 percent in 1990), while the share of fish farming jobs is rising.[35] Some 16.6 million of the world's fishers (30 percent) are engaged in fish farming.[36]

The overwhelming majority of people working in the fisheries sector are in Asia (87 percent), followed by Africa (more than 7 percent), and Latin America and the Caribbean (3.6 percent).[37] Europe, North America, and Oceania together account for only slightly more than 2 percent.[38]

Counting dependents as well as people employed in processing, packaging, marketing, and distribution of fish, those manufacturing fishing equipment (including boats, nets, and other gear), and the people working in research, development, and administration, anywhere from 660 million to 820 million people are economically dependent upon the fishery sector.[39] This is 9–12 percent of the world's population.[40]

Small-scale fishers and fish farmers account for about 90 percent of the world's fishers.[41] But they face a series of handicaps, including a lack of bargaining power, inadequate access to credit, difficulty in meeting market access regulations, and poor trade-related infrastructure.[42]

On average, each person employed in capture fisheries in 2010 was responsible for an annual catch of 2.3 tons, much lower than the 3.6 tons for aquaculture.[43] There are enormous differences between the highly mechanized operations of vessels from Europe (25.7 tons per year of wild and farmed fish per person) and North America (18.2 tons), which trawl huge ocean areas, and the typically smaller-scale, less-mechanized (and even nonmotorized) operations of many fishers in Asia (2.1 tons), Africa (2.3 tons), or even Latin America and the Caribbean (6.9 tons).[44]

The vast majority of the world's fishing vessels are found in Asia. In 2010, it accounted for 3.18 million vessels—73 percent of the global 4.36 million.[45] All other regions have far smaller fleets, with Africa at 11 percent, Latin America and the Caribbean at 8 percent, and North America and Europe each at 3 percent.[46]

The world is still coming to terms with decades of overfishing. The share of stocks that are overexploited or depleted increased from 10 percent in 1974 to 29.9 percent in 2009; the share of fully exploited stocks rose from 51 percent to 57 percent; and the proportion of non-fully-exploited stocks declined from close to 40 percent to just 12.7 percent.[47] Some critics charge that the share of overexploited stocks is in fact far higher than these numbers suggest, but there is no consensus on that point.[48]

As fishing fleets pushed into ever-deeper waters, slow-reproducing deep-sea fish were easily overexploited and habitats damaged. Also, illegal, unreported, and unregulated fishing continues to be widespread, amounting to anywhere from 14 to 33 percent of the world's total catch.[49] A variety of measures have been adopted to address overfishing, such as imposing quotas or a total allowable catch for a given year or fishing season, establishing marine protected areas, or prohibiting dragnet fishing and bottom trawling.[50] There are also efforts to use eco-labels and certification. Marine Stewardship Council certificates, for instance, encourage sustainable fisheries management, but getting one of these is prohibitively expensive for small fishers.[51]

Meanwhile, the massive expansion of aquaculture has triggered environmental and health concerns. These include the growing degradation of land and marine habitats (such as the clearing of mangrove forests to make way for shrimp ponds), pollution of adjacent waters from fertilizer and antibiotics effluents, the spread of diseases among fish populations raised under the crowded conditions of intensive fish farming, and the effects of the overuse of antibiotics and pesticides.[52]

Some efforts are under way to address these problems. These include certification systems like the Aquaculture Stewardship Council (which is trying to play a role in fish farming similar to the Marine Stewardship Council's role with regard to marine capture fisheries) and the Global Aquaculture Alliance's Best Aquaculture Practices.[53]

For both capture and farmed fish, climate change may well compound existing problems and threats.[54] More-sustainable management of marine and coastal habitats is essential, and it would also bolster the resilience of coastal communities, especially small-scale fishers.[55] But fishers also need to adopt more-sustainable practices in aquaculture, and policy makers need to discourage unsustainable and illegal fishing methods.

Peak Meat Production Strains Land and Water Resources

Michael Renner

Global meat production rose to an estimated 308.5 million tons in 2013, an increase of 1.4 percent over 2012.[1] The United Nations Food and Agriculture Organization (FAO) forecast additional growth of 1.1 percent in 2014 to 311.8 million tons.[2] Production is thus reaching new peaks, despite drought conditions in Australia and New Zealand and disease outbreaks in the United States and Eastern Europe.[3] However, the annual rate of growth has slowed from 2.6 percent in 2010.[4]

Since 1961, the first year with comparable FAO data, production has expanded more than fourfold, responding to growing purchasing power, urbanization, and changing diets.[5] (See Figure 1.) Looking back much further, meat production has grown 25-fold since 1800, outpacing human population growth by a factor of 3.6.[6]

Asia's 131.5 million tons accounted for close to 43 percent of world output in 2013.[7] Europe was second, with 58.5 million tons, followed by North America (47.2 million tons) and South America (39.9 million tons).[8] Africa was a distant fifth, at 16.5 million tons, slightly above the combined production of Central America and Oceania.[9]

Among individual countries, China was without equal, producing 85.5 million tons in 2013—close to 28 percent of the world total.[10] (See Figure 2.) European Union (EU) members (44.9 million tons), the United States (42.8 million tons), and Brazil (25.3 million tons) follow, and together with China they accounted for two-thirds of global output.[11] No other country came close, although Russia, India, Mexico, Argentina, Canada, and Australia stood out among the remaining producers.[12]

But in an age of globalization, it is not enough to look at statistics through a national lens. The 10 largest meat companies, measured by their 2011–13 sales,

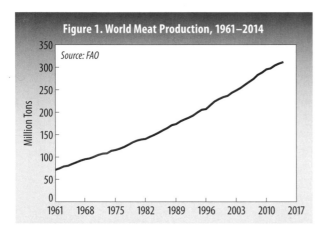

Figure 1. World Meat Production, 1961–2014

Source: FAO

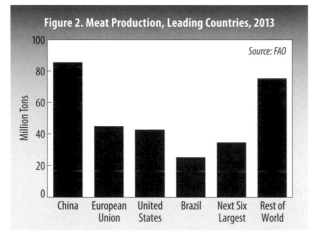

Figure 2. Meat Production, Leading Countries, 2013

Source: FAO

Michael Renner is a senior researcher at Worldwatch Institute and codirector of *State of the World 2015.*

are headquartered in just six countries: JBS (Brazil), Tyson Food (United States), Cargill (United States), BRF (Brazil), Vion (Netherlands), Nippon Meat Packers (Japan), Smithfield Foods (United States, but acquired by China's Shuanghui International Holdings in 2013), Marfrig (Brazil), Danish Crown AmbA (Denmark), and Hormel Foods (United States).[13]

The share of meat that is traded rose from less than 4 percent of total production in 1961 to 10 percent in 2013.[14] The two most important exporters in 2013 were the United States (7.6 million tons) and Brazil (6.4 million tons), together representing 45 percent of global trade.[15] Other important sellers were the EU (4 million tons), Australia (2 million tons), China (1.9 million tons), India (1.8 million tons), and Canada (1.7 million tons).[16]

Importers were less concentrated. China (4.4 million tons), Japan (3.1 million tons), and Russia (2.4 million tons) were the largest buyers and together accounted for one-third of global imports.[17] They were followed by Mexico, the United States, Vietnam, the EU, and Saudi Arabia, which in all accounted for another 24 percent of all imported meat in 2013.[18]

Meat consumption basically follows the same regional pattern as production, with Asia being dominant. On a per capita basis, meat use stood at 42.9 kilograms (kg) in 2013 worldwide.[19] Although the disparities have narrowed somewhat, people in industrial countries continue to eat much larger quantities—75.9 kg—than those in developing nations—33.7 kg.[20] Meat consumption in Japan is much lower than in many other rich countries, however, and runs close to the world average.[21] People in New Zealand (126.7 kg per person) and Australia (121.1 kg) ate the most meat in 2011—far more than those at the bottom of the scale, in Bangladesh and India (just above 4 kg).[22] (See Figure 3.) People in China consume almost 14 times as much meat per person as people in India do, and South Africans consume six times more meat than Nigerians.[23]

The type of meat eaten varies substantially among different countries. For example, 54 percent of the meat consumed by Argentinians is beef, whereas 74

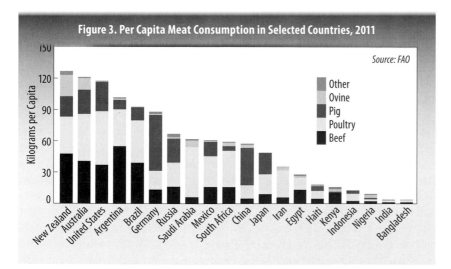

Figure 3. Per Capita Meat Consumption in Selected Countries, 2011

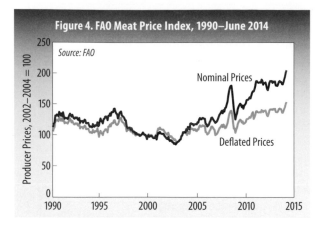

Figure 4. FAO Meat Price Index, 1990–June 2014

Source: FAO

Nominal Prices

Deflated Prices

percent of what Saudis consume is poultry and 61 percent in Germany is pork, while in Nigeria, lamb and mutton (ovine meat) account for 30 percent.[24]

The growth in meat consumption has not been constrained by rising prices. Following a decline in the 1990s, the FAO meat price index has been on the rise over the past decade, going up 2.4-fold in nominal terms and 1.7-fold after inflation.[25] (See Figure 4.) The only exception to this upward trend came in 2008/09, when prices briefly plummeted.[26] During the first half of 2014, in particular, prices rose sharply.[27]

Pork and poultry accounted for 72 percent of global meat production in 2013.[28] At 114.3 million and 107 million tons, respectively, they were far ahead of bovine meat (67.7 million tons) and ovine meat (13.9 million tons).[29] Trade, however presents a different picture. Poultry (13.2 million tons) had a 43-percent share, followed by bovine meat (9.1 million tons), pig meat (7.4 million tons), and ovine meat (1 million tons).[30]

Just a dozen countries are the principal producers, exporters, and importers of each type of meat, with the United States, China, and the EU playing particularly central roles.[31] China single-handedly accounted for 48 percent of global pig meat production in 2013.[32] Just two countries—Australia and New Zealand—were responsible for a stunning 84 percent of the world's lamb and mutton exports.[33]

The steady growth of global meat production and use comes at considerable environmental and health cost. The livestock sector is characterized by industrial methods that use large quantities of water, that mix high-value grains such as corn or soybeans into feed, that cut down forests to expand grazing lands, that confine animals in cramped conditions, that replace organic fertilizers (manure) with synthetic nitrogen, and that administer heavy doses of antibiotics to speed animal growth and reduce the likelihood of disease outbreak.[34]

In the United States, 13,600 tons of antibiotics were sold for use in livestock operations in 2011—almost four times the 3,500 tons used to treat ill people.[35] In Europe, an estimated 8,500 tons were distributed for animal use, but this paled compared with possibly more than 100,000 tons in China.[36]

Livestock use large amounts of land. Close to 70 percent of the planet's agricultural land is used for animal pasture, and another 10 percent is used to grow grains fed to livestock.[37] Producing beef is much more resource-intensive than producing pork or chicken, requiring roughly three to five times as much land to generate the same amount of protein.[38] Beef uses about three-fifths of global farmland but yields less than 5 percent of the world's protein and less than 2 percent of its calories.[39]

Feeding grain to livestock improves their fertility and growth, but it sets up a de facto competition for food between cattle and people. Worldwide, close to 800 million tons of wheat, rye, oats, and corn are fed to animals annually (more than

40 percent of world production), along with 250 million tons of soybeans and other oilseeds.[40]

Meat production also consumes a lot of water. Agriculture uses about 70 percent of the world's available freshwater, and one-third of that—more than 20 percent of all water consumed—is used to grow the grain fed to livestock.[41] Beef is by far the most water-intensive of all meats.[42] (See Table 1.) The more than 15,000 liters of water used per kilogram is far more than is required by a number of staple foods, such as rice (3,400 liters per kg), eggs (3,300 liters), milk (1,000 liters), or potatoes (255 liters).[43]

Alternative practices could reduce these environmental and health impacts, such as switching feed from grains to grass and other plants, using natural instead of synthetic fertilizers, and ending factory-style livestock operations. But dietary choices also make a big difference. Eating less meat typically means leading a less resource-intensive life. What matters, however, is not only how much meat people eat but also the kind of meat they consume.

Table 1. Water Requirements of Different Types of Meat

Meat	Liters/Kilogram	Liters/Gram of Protein
Bovine Meat	15,415	112
Sheep/Goat Meat	8,763	63
Pig Meat	5,988	57
Chicken Meat	4,325	34

Source: Water Footprint, "Animal Products" at www.waterfootprint.org/?page=files/Animal -products.

Coffee Production Near Record Levels, Prices Remain Volatile

Michael Renner

According to the U.S. Department of Agriculture's Foreign Agricultural Service (USDA-FAS), world coffee production during the 2013/14 crop year was just slightly over 9 million tons, down 3.2 percent from the record 9.3 million tons in 2012/13.[1]

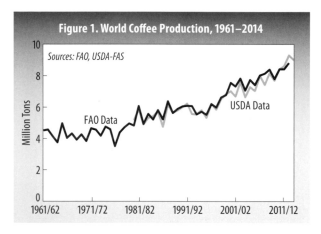

The decline was largely due to developments in Brazil, where production has suffered from a mix of such adverse factors as frost damage in Paraná, prolonged drought, and high temperatures in Minas Gerais and São Paulo states.[2]

Although zig-zagging from year to year, production has been on an overall steady upward trend, especially from the mid-1990s.[3] According to the United Nations Food and Agriculture Organization (FAO), which offers a longer time series, production today is double the 4.5 million tons in 1961/62.[4] FAO does not yet have data for 2013/14, however, and over the last decade, FAO and USDA-FAS have offered somewhat different production estimates.[5] (See Figure 1.)

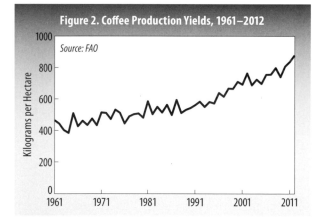

Notwithstanding some ups and downs between the early 1970s and early 1990s, the total area of land devoted to coffee cultivation has stayed within the same general range. In 1961, the global total was 9.8 million hectares (ha), hitting a low of 7.9 million ha in 1976, then a peak of 11.2 million ha in 1990, only to return to 10 million ha by 2012.[6] Gains in output have principally been due to rising yields. Over the last half-century, the average harvest per hectare has improved from 384 kilograms (kg) in 1964 to 879 kg in 2012.[7] (See Figure 2.)

South America is the dominant producer of coffee and accounted for 45 percent of global output in 2012/13, followed by Asia (30 percent), Central America (13 percent), and Africa (11 percent).[8] The regional balance has thus shifted significantly from half a century ago, when South America controlled 65 percent of the total, Africa had 20 percent, but Asia had less than 5 percent.[9]

Michael Renner is a senior researcher at Worldwatch Institute and codirector of *State of the World 2015.*

Brazil is perched atop the producers' league, with its 3.2 million tons accounting for 36 percent of the world's coffee production in 2013/14.[10] Vietnam was in second place with 1.7 million tons (19 percent), followed at a distance by Colombia (660,000 tons, 7 percent), Indonesia (570,000 tons, 6 percent), and Ethiopia (381,000 tons, 4 percent).[11] Together with the next five largest producers (India, Honduras, Peru, Uganda, and Mexico), the top 10 accounted for 87 percent of total production, up from 75 percent in 2000.[12] (See Figure 3.)

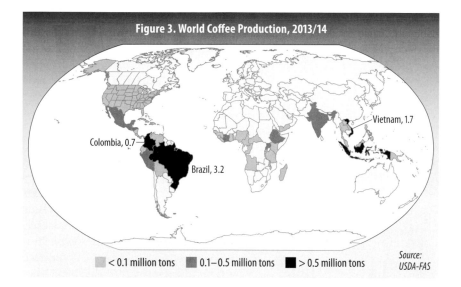

Figure 3. World Coffee Production, 2013/14

Vietnam, 1.7

Colombia, 0.7

Brazil, 3.2

< 0.1 million tons   0.1–0.5 million tons   > 0.5 million tons

Source: USDA-FAS

There are two main types of coffee. Arabica beans are a sweeter and milder type grown in higher altitudes (cultivated mostly in South and Central America), whereas Robustas have a more bitter flavor and are grown in lowlands (mostly in Africa and Asia).[13] Arabica's once-dominant role has diminished, as Asian producers have risen in importance. In 1983/84, Arabica's share peaked at nearly 80 percent, but it has now declined to below 60 percent.[14]

The vast majority of coffee production—between 75 and 80 percent—is exported, flowing from developing countries principally to industrial ones.[15] The largest consumers in 2013/14 were the European Union (2.7 million tons, or 31.5 percent of the world's total), the United States (1.5 million tons, 17.2 percent), and Brazil (1.2 million tons, 13.8 percent), followed at a distance by Japan (459,000 tons, 5.3 percent).[16] Russia, the Philippines, Canada, Ethiopia, Indonesia, and Switzerland are the next-largest consumers; together, they use 1.2 million tons, or 13.6 percent of the global total.[17] The top 10 consumers combined account for 81.3 percent of worldwide demand.[18] (See Figure 4.)

International coffee prices (Arabicas) were on the rise from 2002, when they hit lows not seen in almost 40 years, until early 2011, when prices began to drop sharply.[19] The long-term developments since 1960 demonstrate a continued pattern of tremendous price volatility.[20] (See Figure 5.)

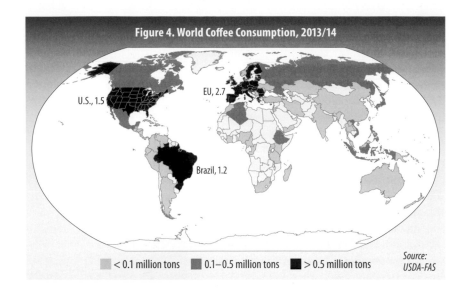

Figure 4. World Coffee Consumption, 2013/14

U.S., 1.5   EU, 2.7   Brazil, 1.2

< 0.1 million tons   0.1–0.5 million tons   > 0.5 million tons

Source: USDA-FAS

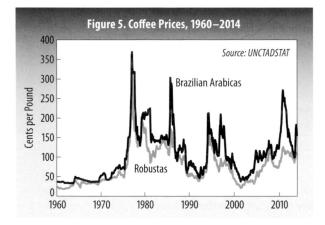

Figure 5. Coffee Prices, 1960–2014

Source: UNCTADSTAT

Brazilian Arabicas

Robustas

There are multiple reasons for coffee price fluctuations, including, most recently, the so-called financialization of commodities, under which money flows from speculative investors (including hedge funds, commodity indexes, and exchange-traded-funds) have surged.[21]

A discussion paper by the U.N. Conference on Trade and Development notes that "structural vulnerabilities in most CDDCs [Commodity Dependent Developing Countries] render their economies more vulnerable to increased commodity market turbulence than developed countries, given their comparatively lower income and high dependence on commodity exports."[22]

Because a number of developing countries depend heavily on commodities like coffee for the bulk of their export earnings, extreme price fluctuations have destabilizing impacts on their economies. This is especially the case for a number of African countries. In Burundi, coffee revenues represent 79 percent of foreign exchange earnings, and in Ethiopia, 64 percent.[23]

Extreme price volatility also threatens the livelihoods of many small producers. An estimated 25 million people worldwide cultivate this labor-intensive crop, but many are small-scale producers who derive limited economic benefit compared with the traders, roasters, distributors, retailers, and investors who often capture the bulk of revenues and have bigger influence over market developments.[24] World market prices are set far from the places where coffee is grown—on the Intercontinental Exchange in New York (Arabicas) and the London International Financial Futures Exchange (Robustas).[25]

The social and environmental conditions under which coffee is produced differ widely among countries. Concerns are far-ranging, including work conditions and workers' rights (wages, collective bargaining, occupational health and safety, forced labor and child labor, etc.), as well as deforestation, impacts on biodiversity, and the use of agrochemicals. Prices for organic coffee tend to be more stable than those for conventional coffee, and access to organic fertilizers is cheaper. Fair trade cooperatives typically provide purchase guarantees, technical assistance, and access to credit.[26]

Multiple initiatives have emerged that seek to define and implement more-sustainable practices.[27] The key standards are fair trade (guaranteeing a minimum price for smallholders), organic (replacing inorganic fertilizers, pesticides, and fungicides with organic ones, and often requiring coffee to be shade-grown), Utz Certified (incorporating a set of social and environmental criteria), and Rainforest Alliance (based on integrated pest management principles and including provisions for worker welfare).[28] There is also the Common Code for the Coffee Community, which uses the most basic standard of all initiatives, and Starbucks and Nespresso maintain their own private company standards.[29]

In 2009, it was estimated that coffees certified as sustainable under various initiatives represented 8 percent of worldwide coffee exports.[30] Certified coffee is growing rapidly, however, and at recent growth rates could conceivably reach 20–25 percent of global coffee trade by 2015.[31] In the Netherlands, 40 percent of all coffee is certified; in the United States the figure is 16 percent; the share in Scandinavian countries is more than 10 percent; Europe's largest market, Germany, has a much lower 5-percent certified rate; and there are generally low percentages in southern Europe.[32] Demand for certified coffee is also growing in urban areas of countries like China, India, Mexico, and Brazil.[33]

Cooperatives play an important role in empowering rural communities—improving their access to knowledge, inputs, and finance, as well as providing fair market access.[34] In Ethiopia, a successful example is the Oromia Coffee Farmers Cooperative Union (OCFCU), operating in Oromia Regional State, an area that accounts for more than two-thirds of the country's total coffee-growing land.[35] OCFCU has grown from an initial 22,691 household farmers in 1999 to 254,052 farmers at present.[36] OCFCU has its own coffee processing facility, employing 1,200 people seasonally; has provided important educational, health, and other benefits for its members; and has created a bank providing crucial preharvest financing and offering insurance against crop loss.[37]

In Costa Rica and India, cooperatives have become leaders in the production of carbon-neutral coffee.[38] The coffee sector's own contribution to carbon emissions is small compared with other sources. But growers will need to pay increasing attention to climate change. Rising temperatures, altered rainfall patterns, and rising pest incidence will increasingly affect future coffee production, requiring adaptation measures.

Volatile Cotton Sector Struggles to Balance Costs and Benefits

Michael Renner

Cotton has been grown by human civilizations for thousands of years. Promoted as "the fabric of our lives" in a long-running U.S. advertising campaign, it remains by far the most important natural fiber for textiles, although it continues to lose ground to synthetic fibers. Cotton's share of all fibers worldwide was 88 percent in the 1940s and 68 percent in 1960.[1] But by 2010, it had declined to just 33 percent, while synthetic fibers grew to 60 percent, and wool, flax, and cellulosic fibers accounted for the remainder.[2]

Growing cotton provides livelihoods for an estimated 100 million households in as many as 85 countries.[3] But adverse global market conditions, volatility, and reliance on large doses of water, fertilizer, and pesticides impose considerable social and environmental costs.

In 2013/14, an estimated 26.3 million tons of cotton were produced worldwide, 5.5 percent below the peak value of 27.8 million tons in 2012.[4] Production has expanded 3.7-fold since 1949/50, outstripping the 2.9-fold growth in human numbers.[5] Exports have typically accounted for between 27 and 38 percent of total production in any given year.[6] Consumption trends have been less erratic than production, but demand is now down 12 percent from the peak in 2007/08 (see Figure 1)—reflecting the impacts of the world economic crisis and the ongoing replacement of cotton by other fibers.[7]

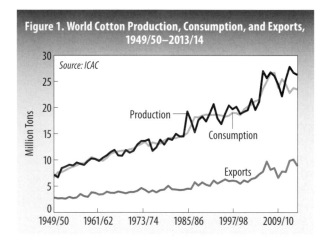

Figure 1. World Cotton Production, Consumption, and Exports, 1949/50–2013/14

Source: ICAC

The global area devoted to cotton has zigzagged over the decades within a relatively narrow band of 30–35 million hectares, which is about 2–3 percent of the world's total arable land.[8] (See Figure 2.) Output growth has instead been driven by rising yields, which have more than tripled from 250 kilograms (kg) per hectare in 1949/50 to 804 kg today.[9] Currently, the highest yields are being achieved by growers in Australia (2,270 kg), Israel (1,809 kg), and Turkey (1,686 kg), followed by Mexico (1,625 kg), Brazil (1,520 kg), and China (1,474 kg).[10]

China and India together account for 52 percent of global output—up from just 13 percent in 1949/50, when the United States produced close to half of all cotton.[11] The top seven producers—which includes Pakistan, Brazil, Uzbekistan, and Turkey—represent 83 percent of the world's current total.[12]

Michael Renner is a senior researcher at Worldwatch Institute and codirector of *State of the World 2015*.

Cotton production tends to vary widely from year to year, influenced by weather conditions, water availability, yields, and cultivation area. China's production, for example, has fluctuated enormously, but since the early 1980s, it has been larger than any other country's.[13] (See Figure 3.) The area used for cotton cultivation in China has declined strongly for a number of reasons—including price weakness, competing demands of food production, and cotton farmers' desire to diversify their crops—and adverse weather conditions depressed yields in some years.[14] Following a steep rise in the last decade, India's output now rivals that of China and is forecast to surpass it during the next two years.[15]

A very similar group of countries dominates consumption. China and India together accounted for 32 percent of world consumption in 2013/14 (slightly less than their share of the world population), and the top seven countries account for three-quarters of the total.[16] China's cotton consumption multiplied 2.4-fold between 1998/99 and 2009/10 but then dropped precipitously.[17] (See Figure 4.) Behind this development was the government's increase of minimum support prices and stock-building, which raised domestic cotton prices substantially but lowered demand.[18] India's cotton demand grew less rapidly but more steadily.[19] U.S. consumption fell from the late 1990s, whereas export-driven demand for textiles drove up cotton use in Pakistan and Bangladesh.[20]

Countries in West and Central Africa—especially Burkina Faso, Mali, Côte d'Ivoire, and Benin—are relatively small producers (with a joint volume comparable to that of Uzbekistan, the sixth-largest producer), but because most of their crop is sold abroad, they account for more than 10 percent of global exports (behind the United States, India, and Australia).[21]

Cotton accounts for 5–10 percent of the gross domestic product of these countries, but this metric understates the extent of their dependence on this single crop.[22] Mali derives 76 percent of its export earnings from cotton, and 40 percent of that country's rural population depends on the crop

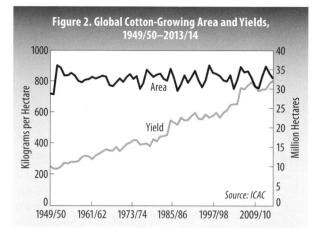

Figure 2. Global Cotton-Growing Area and Yields, 1949/50–2013/14

Source: ICAC

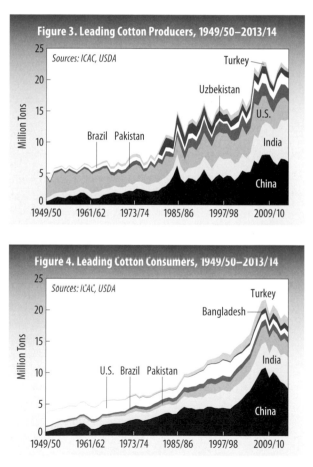

Figure 3. Leading Cotton Producers, 1949/50–2013/14

Sources: ICAC, USDA

Figure 4. Leading Cotton Consumers, 1949/50–2013/14

Sources: ICAC, USDA

for their livelihoods.[23] In Benin, cotton provides 86 percent of foreign earnings, and in Burkina Faso 68 percent.[24] This compares to 38 percent in Uzbekistan, and just 8 percent in India and 7 percent in Pakistan.[25]

Except for a sharp but brief spike in late 2010 and early 2011, world market prices have in most years fluctuated within a narrow band of 50¢–100¢ per pound.[26] But adjusted for inflation, the International Cotton Advisory Committee reports that prices have been on a downward trajectory, to as little as one-quarter of their 1950s level, offsetting the growth in output over the last five decades.[27]

Declining cotton prices are the result of a number of factors, including substitution by synthetic fibers, stocks that have risen to more than 80 percent of annual consumption, and subsidies that allow some producers to dump their output below cost.[28] According to a Fairtrade Foundation report, the United States, China, and the European Union together spent $47 billion on subsidies between 2001 and 2010.[29] In 2009/10, China overtook the United States as the largest provider of subsidies; in 2012, it spent an estimated $3 billion, compared with $820 million in the United States.[30]

The legions of small cotton farmers in the world face a set of challenges largely beyond their control. In addition to unfair subsidies and declining and volatile prices, they must deal with health risks from pesticide use and in some cases insurmountable levels of debt.

Cotton is a very pesticide-intensive crop. About $2.6 billion worth of pesticides are used on cotton each year.[31] Growing cotton accounts for 16 percent of global insecticide use and 6.8 percent of herbicide use.[32] Pest resistance is one repercussion, and adverse health impacts on farmers—from acute poisoning to long-term effects—are another.[33] Health effects depend on the type of pesticide. Some affect the nervous system, others irritate the skin or eyes, cause cancer, or affect the hormone or endocrine system in the body.[34] Pesticides and fertilizer (nitrogen, phosphorus, potash) can also leach out of the plant's root zone and contaminate groundwater and surface water.

Sadly, severe indebtedness has caused an estimated 100,000 cotton farmers in India to commit suicide over a 10-year period when they were overwhelmed by their situations.[35] Indebtedness results from a number of factors, including the rising cost of pesticides and genetically modified seeds, low yields due to droughts, and the declining price that cotton fetches on world markets.

Cultivating cotton takes about 3 percent of all agricultural water use worldwide—slightly more than the 2–3-percent share of arable land used.[36] Where cotton is cultivated intensively, it requires large amounts of water. Inefficient irrigation practices cause soil salinization and a decline in soil fertility. In Central Asia, massive amounts of water from the Amu Darya and Syr Darya rivers have been diverted since the 1960s to cultivate cotton and other crops, triggering a catastrophic shrinkage of the Aral Sea.[37]

Some countries, such as Brazil, India, and Mali, rely on rainwater to a high degree.[38] But others, including Australia, Pakistan, Syria, Turkey, Turkmenistan, and Uzbekistan, rely heavily on irrigation—and Egypt does so exclusively, which can be a problematic practice if water withdrawal is unsustainable.[39]

Studies from the 1997–2001 period indicate that cotton grown in China had the lowest "water footprint"—about 2,000 cubic meters (m³) per ton.[40] The United States, Australia, Greece, Mexico, and Brazil follow with values between 2,250 and 2,620 m³/ton.[41] Turkey is slightly below the world average of about 3,600 m³/ton, while Egypt, Uzbekistan, and Pakistan are in the range of 4,200 to 4,900.[42] Cotton from India (more than 8,600), Turkmenistan (6,000), and Mali (5,200) is far more water-intensive.[43]

The efficiency of water use in cotton cultivation keeps improving. Global water use for this purpose was estimated at 256 billion m³ per year during 1997–2001.[44] But the annual figure dropped to 207 billion m³ for 1996–2005.[45] National statistics confirm this picture. For example, U.S. cotton growers reduced the amount of irrigation water applied per unit of production by 75 percent between 1980 and 2011.[46] According to Cotton Australia, that country's growers have improved their water use efficiency by 3–4 percent a year in the last decade.[47]

If all the world's cotton were grown as efficiently as by the top 10 percent of cultivators, 54 percent of current global water use could be avoided.[48] Even if the least-efficient cultivators matched the global average in efficiency, savings would still amount to 30 percent.[49]

Countries importing cotton or finished cotton products bring large amounts of embedded "virtual water" into their nations. The resulting water footprint can be considerable. It has been estimated that producing a pair of jeans takes 10,850 liters of water and a T-shirt takes 2,720 liters.[50]

Several initiatives exist to improve the social and environmental conditions under which cotton is produced. They revolve around internationally recognized organic farming standards and the principle of fair trade. Fair trade standards also often include environmental criteria, restricting the use of agrochemicals and pesticides.[51] Organic production replaces synthetic fertilizers and pesticides with organic substances and integrated pest management, relies on soil fertility management, and entails a number of harvest and post-harvest quality management measures.[52]

Fair trade labels for cotton have been in existence for about a decade. Producers participating in them are usually small family farms organized in cooperatives or associations.[53] Under such labels, cotton farmers receive a minimum price covering the average costs of sustainable production, as well as a premium.[54] Currently, about 1.5 million farmers (affecting 7.5 million people when their families are included) are involved in the cotton fair trade system in Latin America, Africa, and Asia.[55]

Another effort is found in the Better Cotton Initiative (BCI), supported by a number of retailers and active in 19 countries. BCI seeks to reduce the environmental impact of cotton production, improve the livelihoods of farmers, and promote decent work.[56] For instance, it helped Pakistani farmers reduce pesticide use by 47 percent and chemical fertilizer use by 39 percent in 2012; yields were maintained and incomes improved 11 percent relative to farmers who stuck with conventional practices.[57] In 2013, just 3.7 percent of all cotton was produced in accordance with BCI principles, but the goal for 2020 is to cover 30 percent of cotton and involve 5 million farmers.[58]

These initiatives offer important benefits to cotton farmers. For the moment, at least, they still account for a relatively small share of the entire industry.

Genetically Modified Crop Industry Continues to Expand

Wanqing Zhou

Genetically modified (GM) crops have had their genetic materials engineered through biotechnologies to introduce new or enhanced characteristics (traits), such as herbicide tolerance, insect resistance, enhancement of certain nutrients, and drought tolerance. The global plantation area of GM crops has been growing for more than two decades, since they were first commercialized in the early 1990s, and it reached 181.5 million hectares in 2014.[1] (See Figure 1.) But the annual growth rate has slowed considerably, from over 125 percent in the late 1990s to 6.3 percent in the early 2010s.[2]

According to the International Service for the Acquisition of Agri-biotech Applications, an organization that has been keeping track of official information on GM crop field trials and plantings, North America and South America accounted for 87 percent of the global GM crop area in 2014, with 84.7 million and 73.3 million hectares, respectively.[3] They were followed by Asia (19.5 million hectares), Africa (3.3 million hectares), Oceania (0.5 million hectares), and Europe (0.1 million hectares).[4]

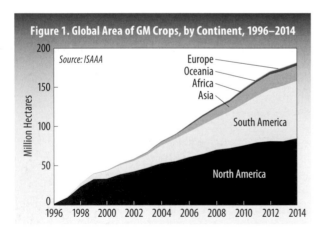

Figure 1. Global Area of GM Crops, by Continent, 1996–2014

The peak growth in global planted area occurred in 1998.[5] The 16.8-million-hectare increase that year—nearly twice the increase in 1997—was largely attributed to the rapid expansion of GM crops in the United States and Argentina.[6] In the first decade of the twenty-first century, active growth in planted area expanded from North and South America to Asia and Africa. For most of the 2010s, developing countries in South America (mainly Brazil and Argentina), Asia (mainly China and India), and Africa (mainly South Africa) added a larger area than the industrial countries did. Since 2012, the developing world has been planting larger areas of GM crops than industrial countries have. (See Figure 2.) However, the growth rates in both the industrial and the developing world are declining.

In 2014, among the 18 million farmers from the 28 GM-crop-planting countries, 1.5 million lived in eight industrial countries, led by the United States.[7] But 90 percent of the farmers lived in 20 developing countries, which accounted for 53 percent of the total planted area.[8] The data also reveal that planting GM crops is 9.8 times as labor-intensive in developing countries—a differential that has narrowed from 14 times as labor-intensive in the early 2000s.

Wanqing Zhou is a research associate in the Food and Agriculture Program at Worldwatch Institute.

The top 15 countries together grew 99.7 percent of GM crops in 2014. The largest 5 producers alone planted 89.7 percent of all GM crop areas on the planet. (See Table 1.) Interestingly, 6 of the 15 largest producers—all 6 of which are developing countries—started officially planting GM crops only within the last decade. For example, Brazil did not officially approve the planting of GM crops until 2003. But double-digit annual growth rates in almost all the years since then lifted Brazil into second place worldwide. Its GM crop now mainly consists of herbicide-tolerant soybeans.[9]

In 2014, the most widely grown GM crops were soybeans (90.7 million hectares), maize (55.2 million hectares), cotton (25.1 million hectares), and canola (9 million hectares).[10] Other commercially planted GM crops include sugar beet, alfalfa, papaya, potato, squash, tomato, sweet pepper, and eggplant.[11] Insect-resistant poplar trees have been commercially planted in China.[12] GM loblolly pines with increased wood density are also approved for unregulated release in the United States.[13]

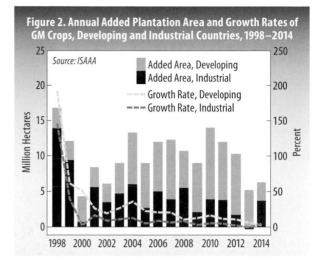

Figure 2. Annual Added Plantation Area and Growth Rates of GM Crops, Developing and Industrial Countries, 1998–2014

Table 1. Area Planted in GM Crops, by Country and Share of Global Total, 2014

Country	Area Planted	Share of Global Total
	(million hectares)	(percent)
United States	73.1	40.3
Brazil	42.2	23.3
Argentina	24.3	13.4
India	11.6	6.4
Canada	11.6	6.4
China	3.9	2.1
Paraguay	3.9	2.1
Pakistan	2.9	1.6
South Africa	2.7	1.5
Uruguay	1.6	0.9
Bolivia	1.0	0.6
Others	2.7	1.5

Source: C. James, Global Status of Commercialized Biotech/GM Crops: 2014 (Ithaca, NY: ISAAA, 2014).

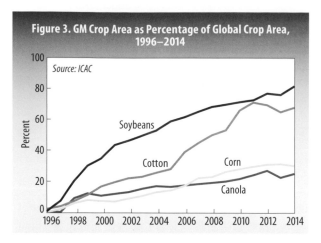

Figure 3. GM Crop Area as Percentage of Global Crop Area, 1996–2014

Source: ICAC

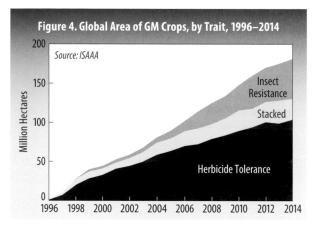

Figure 4. Global Area of GM Crops, by Trait, 1996–2014

Source: ISAAA

Steady growth in the planted area for the four major GM crops during 2000–10 has given way to a more fluctuating pattern in the early 2010s. GM cotton experienced a moderate decline in 2012 and 2013, mainly due to lower prices.[14] The plantation area of GM canola dropped by 1 million hectares in 2013, as Canada increased the share of non-GM wheat planted in the rotation of crops.[15] These developments, together with the prominent increase in global non-GM soybean plantation area in 2013, resulted in declines in the percentages of GM soybean, cotton, and canola in global areas that year.[16] (See Figure 3.)

The slower growth of GM crop plantation area is due to the saturation of adoption rates in countries with "mature" GM crop business models.[17] For example, adoption rates have exceeded 90 percent for major GM crops in the United States. And 95 percent of canola in Canada, 99 percent of cotton in Australia, 95 percent of cotton in India, 99 percent of cotton and 100 percent of soybean in Argentina, and over 95 percent of soybean in Paraguay are already genetically modified.[18]

Herbicide tolerance has been the dominant trait among all officially approved GM crops since 1996, accounting for nearly 60 percent of all commercial GM crops—mainly soybean, canola, and maize—in 2014.[19] Stacked traits (which include different types of insect resistance or combine insect resistance and herbicide tolerance in a single crop variety) surpassed modification for insect resistance on its own in 2007. Crops with stacked traits covered 51 million hectares, or 28 percent of GM crop areas globally, in 2014.[20] (See Figure 4.) Stacked traits are often used in maize and cotton. But stacked soybean has also been commercially planted in Brazil since 2013. It reached 5.2 million hectares in 2014, more than double the previous year.[21] Other traits introduced include virus resistance, drought tolerance, nutrient enhancement, and market value enhancement.[22]

The global value of GM seed reached $15.7 billion in 2014, up 0.6 percent from 2013—a growth rate unprecedentedly low compared with double-digit annual growth rates through the first decade of the twenty-first century.[23] A small handful of companies that develop and market GM crops have a near-monopoly. In the United States, for example, Monsanto holds 63 percent of all the Release Permits and Release Notifications issued by the U.S. Department of Agriculture (USDA) so far, and DuPont Pioneer holds another 13 percent, followed by Syngenta (6 percent), Dow AgroSciences (4 percent), USDA (4 percent), and smaller companies.[24]

One of the familiar narratives for the promotion of GM crops was that it could help alleviate poverty and hunger, but the real effect deserves closer assessment. Instead of producing more food by improving yield—a common perception of how GM crops can be used to reduce hunger—the benefit of these technologies more often consists of saving time and effort in farming, as well as reduced market risks for farmers.[25] The current profile of GM crops clearly shows that the principal driving force is demand for animal feed (soybean and maize) and oil (soybean and canola) instead of for food directly consumed by people.

This raises at least two noteworthy concerns. First, from a social perspective, although the efficiency improvement may give farmers time to turn to other sources of income, it has led to loss of land and livelihoods when farmers with more assets expand to take over the land of less resourceful and less protected small farmers.[26]

Second, from an environmental perspective, the high and growing demand for meat and other animal products, largely met by industrial production methods, is by itself the cause of numerous environmental problems—from pollution to deforestation. In terms of pesticide use and tillage requirement, growing herbicide-tolerant soybean and maize might be less damaging than conventional ways of meeting the demand for animal feed, but the advantage is diminishing as herbicide resistance develops in weeds.[27]

In the next 5–10 years, the profile of commercial GM crops may diversify in both crop variety and traits, based on a list of 71 GM crops that have undergone field trials.[28] These include fruits, protein seeds, and staple foods such as rice and cassava.[29] The broadening of GM crop varieties will likely create more space for growth in GM plantings in the near future, which will require rigorous regulatory frameworks based on the principle of case-by-case assessment.

Food Trade and Self-Sufficiency

Gary Gardner

Imports of grain globally increased more than fivefold between 1960 and 2013 as more nations turned to international markets to help meet domestic food demand.[1] (See Figure 1.) For some countries, the imported share of domestic grain consumption has risen substantially.[2] In 2013, more than one-third of the world's nations—77 in all—imported at least 25 percent of the major grains they needed.[3] This compares to just 49 countries in 1961, an increase of 57 percent over half a century.[4] (See Table 1.)

Even more worrying, 51 countries—about one-quarter of the community of nations—imported more than half of their grain in 2013, and 13 imported all of the grain they needed.[5] Meanwhile, the number of grain-exporting countries expanded by just 6 between 1961 and 2013.[6]

Determining the food import dependence of people, rather than of countries, is more challenging, because imported food is often consumed in a few locations (such as a country's capital city) rather than distributed equally among an entire population. But a 2013 study found that in 2000, some 950 million people—16 percent of the world's population at the time—were using international trade to meet their food needs (although not just grain).[7] The major grains—corn, wheat, and rice—are considered a proxy for food overall, because a large share of the calories consumed by most people comes from grain either directly or indirectly in the form of meat, milk, cheese, and other livestock products.

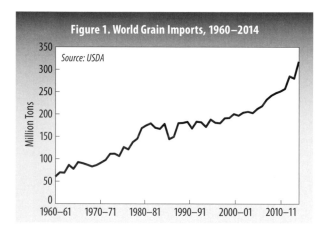

Figure 1. World Grain Imports, 1960–2014

Source: USDA

Among developing countries, grain import dependence is greater than 50 percent in Central America, where land is relatively scarce, and in the Middle East and North Africa, where water is the chief constraint.[8] (See Figure 2.) Sub-Saharan Africa imports about 20 percent of its grain, and the low- and middle-income nations of Asia import about 7 percent.[9] On the other hand, Japan—with the wealth to outbid other nations in international markets—imports about 70 percent of its grain.[10]

Food import dependence has several roots. One problem is the steady loss of fertile land and freshwater. A 2013 study concluded that in 22 countries the consumption of agricultural products (not just grains) requires more freshwater

Gary Gardner is the director of publications at Worldwatch Institute and codirector of *State of the World 2015.*

Table 1. Number of Grain-Importing and Grain-Exporting Countries, 1961 and 2013			
	1961	2013	Increase
	(number of countries)		(percent)
Grain-Importing Countries			
100 percent dependent	11	13	18
More than 50 percent dependent	31	51	65
More than 25 percent dependent	49	77	57
Grain-Exporting Countries	21	27	29

Source: USDA

than each country can extract, and in 62 countries the area of farmland is insufficient to meet domestic consumption needs.[11]

Already, importing food is a de facto water management strategy for many water-scarce countries, because their water supply problem can be shifted to exporting nations. Agriculture commands upward of two-thirds of water withdrawals in most economies, so a smaller agricultural sector can save a great deal of water.[12]

The water embodied in the production of goods and commodities, known as "virtual water," gives a sense of the water savings possible through trade. At the global level, most

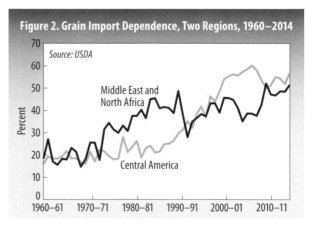

Figure 2. Grain Import Dependence, Two Regions, 1960–2014

embodied or virtual water is associated with agricultural output: some 76 percent of the global flows of virtual water is captured in crops, and another 12 percent is embodied in livestock products.[13]

The biggest net exporters of virtual water are the United States, Canada, Brazil, Argentina, India, Pakistan, Indonesia, Thailand, and Australia.[14] The biggest net importers are North Africa and the Middle East, Mexico, Europe, Japan, and South Korea.[15] Jordan, for example, imports virtual water that is equivalent to five times its own yearly renewable water resources.[16] At the same time, some countries with high external water dependence, like the United Kingdom and the Netherlands, are not actually water-scace.[17]

Pressure on water supplies for agriculture is becoming widespread. A 2012 study in the journal *Nature* estimated that some 20 percent of the world's aquifers are pumped faster than they are recharged by rainfall, often in key food-producing areas such as the Central Valley and High Plains of the United States, the North China Plain, the Nile Delta of Egypt, and the Upper Ganges of India and Pakistan.[18] And a 2002–09 study of satellite data revealed that the region encompassing the

Tigris and Euphrates river basins had lost 144 cubic kilometers of freshwater, nearly as much as in the Dead Sea, largely from overpumping of aquifers.[19] Similar depletions have been monitored by the satellite program in India, North China, North Africa, Southern Europe, and the United States.[20]

Meanwhile, another 2012 study analyzing scarcity in 405 river basins that contain 75 percent of the world's irrigated area documented severe water scarcity for at least one month a year in 201 of them (and somewhat less scarcity in other months).[21] In 35 of the basins, severe water scarcity is the norm for at least half of the year.[22]

Regarding farmland, the United Nations Food and Agriculture Organization (FAO) reports that essentially no additional suitable farmland remains in a belt around much of the middle of the planet, including countries in the Near East and North Africa, South Asia, and Central America and the Caribbean, many of which still have growing populations.[23] Additional available land is found primarily in South America and Africa, but much of it is of high ecological value (for carbon sequestration or biodiversity conservation, for example) or of marginal quality.[24]

Despite the importance of remaining farmland, land continues to be degraded or paved over on all continents. In 2011, FAO reported that 25 percent of land worldwide was highly degraded and another 8 percent was moderately degraded.[25] Meanwhile, farmland near cities is regularly converted to accommodate housing, industry, and other urban needs. The United States, for example, lost 9.3 million hectares of agricultural land to development—an area the size of the state of Indiana—between 1982 and 2007.[26]

Recently, another threat to national endowments of farmland has emerged in the practice of "land grabbing"—the purchase or leasing of land overseas by investment firms, biofuel producers, large-scale farming operations, and governments. Since 2000, agreements have been concluded for foreign entities to purchase or lease more than 42 million hectares, an area about the size of Japan.[27] About half of this area is intended for use in agriculture, while 25 percent is intended for a mix of uses, some of which is agriculture.[28] (Most of the remaining area is to be used for forestry.) Another nearly 9.8 million hectares are under negotiation.[29] The bulk of the grabbed land is located in Africa, with Asia the next most common region for acquisitions.[30] (See Table 2.)

The largest source of land grabbing is the United States, where investors see an opportunity to make money on an increasingly limited resource.[31] (See Table 3.) Target countries are often land-rich or water-rich—indeed, some land is acquired as much for its access to water as for the land itself. Indonesia, for example, is a water-rich nation that is among the most targeted countries for foreign acquisitions.[32] In addition, contracts often do not take into account the interests of smallholders, who may have been working the acquired land over a long period.[33]

Land grabbing surged from 2005 to 2009 in response to a food price crisis, according to a 2012 report from the Land Matrix.[34] Demand for biofuels is another driver. The 2007 Energy and Independence Security Act in the United States called for a fourfold increase in biofuel production by 2022, while a 2009 European Union directive had a similar stimulative impact.[35] In addition, droughts in the United States, Argentina, and Australia drove interest in land overseas.[36]

Table 2. Land Grabbed by Foreign Entities, by Region, Since 2000

Region	Countries with Grabbed Land	Grabbed Area	Share of Global Grabbed Land
	(number)	(million hectares)	(percent)
Africa	40	20.9	49.5
Asia	16	10.0	23.7
Oceania	1	0.0	0.1
Latin America	16	7.3	17.3
Europe	7	4.0	9.6
Total	80	42.2	100.0

Source: Land Matrix, electronic database, viewed 26 February 2015.

Table 3. Leading Investor and Target Countries for Land Investments Since 2000

Investor Countries	Area Acquired	Target Countries	Area Acquired
	(million hectares)		(million hectares)
United States	7.6	Papua New Guinea	3.8
Malaysia	3.6	Indonesia	3.6
Singapore	2.9	South Sudan	3.5
United Arab Emirates	2.8	Dem. Rep. of Congo	2.8
United Kingdom	2.4	Mozambique	2.2
India	2.1	Congo	2.1
Canada	2.1	Ukraine	2.1
Russian Federation	1.6	Russian Federation	1.7
Saudi Arabia	1.6	Liberia	1.3
Hong Kong (China)	1.4	Sudan	1.3

Source: Land Matrix, electronic database, viewed 26 February 2015.

Importing food as a response to resource scarcity has two clear pitfalls. First, not all countries can be net food importers; at some point the demand for imported food could exceed the capacity to supply it. Indeed, many major supplier regions are themselves experiencing resource constraints, as in the United States, where prolonged drought in California led to the fallowing of nearly 5 percent of that state's agricultual land in 2014.[37]

Second, excessive dependence on imports leaves a country vulnerable to supply interruptions, whether for natural reasons (such as drought or pest infestation in supplier countries) or political manipulation. An import strategy may now be unavoidable for some nations, but it should be considered only reluctantly by countries that can meet their food needs in more conventional ways. A better strategy may be vigilance in conserving agricultural resources wherever possible.[38]

Global Economy and Resources Trends

Samples in a plastics factory, Ningbo, China

For additional global economy and resources trends, go to vitalsigns.worldwatch.org.

Global Economy Remained a Mixed Bag in 2013

Mark Konold and Jacqueline Espinal

The global economy grew at 4.49 percent in 2013, and the gross world product reached $87 trillion.[1] For the fourth year in a row, the growth rate was slower than during the preceding year.[2] Growth was affected by numerous adjustments in macroeconomic policies, by high unemployment rates, and by weak aggregate demand in the majority of the industrial economies in the Organisation for Economic Co-operation and Development (OECD).[3] By contrast, emerging and developing markets continued to experience higher growth rates due to supportive policies. In fact, in 2013 emerging markets accounted for 50 percent of total global growth.[4] (See Figure 1.)

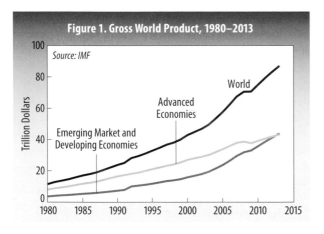

Figure 1. Gross World Product, 1980–2013

Source: IMF

Economic prosperity is often reduced to the lens of growth in gross domestic product (GDP). This is considered the most widely applicable metric for gauging progress and well-being. The gross world product is the sum of the GDPs of all countries. This typically includes levels of consumption, investment, government spending, the cost of imports, and the proceeds from exports.[5] Because of various transaction costs, floating exchange rates, and barriers such as tariffs, economists apply a metric to put purchasing power for countries on an even footing. This metric, applied to the figures in this article, is called the purchasing power parity exchange rate.[6]

The International Monetary Fund forecast growth among developing economies to rise from 4.7 percent in 2013 to 5 percent in 2014 and then to 5.25 percent in 2015.[7] This growth can be traced to the rise of an affluent middle class and a rapid migration of young workers to cities, which encourages more business investment in developing countries.[8] According to the United Nations, Asia and other emerging economies will account for two-thirds of the approximately 370 million people who will have moved to cities by 2015.[9]

Conversely, advanced industrial economies continued to struggle in 2013. Lagging economic growth, especially in the euro area, led to output levels below full capacity, resulting in low inflation and a 7-percent decline in GDP growth rates, down from 3.2 percent in 2012 to 3.0 percent in 2013.[10] As overproduction continued to grow, real interest rates in the private and public sectors are likely to rise, decreasing demand.[11] In the United States, Baby Boomers—individuals born between 1945 and 1965—continued to retire at an approximate rate of 10,000 per

Mark Konold is Caribbean program manager in the Climate and Energy Program at Worldwatch Institute. **Jacqueline Espinal** is an intern in the Climate and Energy Program.

day.[12] In the latter part of the twentieth century, this group contributed strongly to economic growth by spending at record levels and saving less than previous generations did.[13] It is expected that, in retirement, Boomers will reduce their levels of disposable income. Such a change could decrease economic growth by as much as 0.7 percent.[14] Similarly, in Japan, GDP growth between 2000 and 2013 shrank by 0.6 percentage points annually due to an aging population retiring from the workforce.[15] Germany's senior population is also contributing to declining growth by approximately 0.5 percent of GDP.[16] But Germany is, in addition, affected by the rise of a large low-wage sector.[17]

Although the global economy improved following the financial crisis, its growth has not been sufficient to fix the major labor market imbalances created during the crisis.[18] These imbalances include labor shortages due to retirees, increased globalization and competition for available jobs, and a mismatch between current skill levels and job requirements.[19] Active labor market policies (ALMPs)—policies designed to regulate supply and demand of labor, including labor market integration measures offered to the unemployed—have not been effectively implemented.[20] Nearly 202 million people worldwide were unemployed in 2013, a 6-percent unemployment rate.[21] It is estimated that if 1.2 percent of the GDP of industrial economies were allocated to ALMPs, 3.9 million additional jobs could be created in these regions.[22] Yet the dormant effort to implement such policies led 23 million discouraged people to drop out of the labor market.[23]

Worldwide, employment rates declined in all regions except South and East Asia, which continued to experience higher levels of growth through 2013.[24] (See Figure 2.) Although employment rates improved in the United States that year, much of the improvement is attributed to fewer people participating in the labor force—mainly newly retired Baby Boomers.[25] Developing countries, by contrast, were faced with a growing pool of willing laborers in 2013. But limited access to credit for many small enterprises contributed to a lack of investment and job creation in these markets.[26] In 2013, unemployment among people worldwide between the ages of 15 and 24 reached 13.1 percent, up from 11.6 percent in

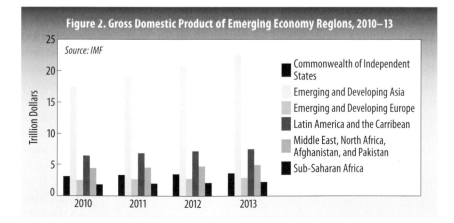

Figure 2. Gross Domestic Product of Emerging Economy Regions, 2010–13

2007.[27] Unfortunately, the number of jobseekers is expected to rise to more than 215 million by 2018, exacerbating the problem.[28]

As the global economy slowly picks up, increased inequality is affecting the development of micro- and macroeconomic policies.[29] For instance, although the United States regained the 8.7 million jobs lost in the recession, wages have declined and the income gap between the rich and the poor has widened to favor the highest-earning 20 percent of households.[30] According to the United Nations Development Programme, average household income inequality in recent decades has risen in both industrial and developing countries.[31] (See Table 1.) This is based on observing what economists call the Gini coefficient, a measure of how income or consumption expenditures among individuals or households within an economy deviate from a perfectly equal distribution. On a scale of 0 to 100, the higher the number, the higher the inequality.[32]

Table 1. Average Gini Coefficient for Various Country Groups and Regions			
	1990s	**2000s**	**Percent Change**
OECD	40.8	44.1	−8.1
Africa	50.7	45.7	9.9
Arab States	38.0	37.9	0.3
Asia & Pacific	40.3	41.7	−3.5
Latin America and Caribbean	47.5	48.0	−1.0
Eastern Europe and Commonwealth of Independent States	32.3	41.2	−27.6

Source: UNDP, Humanity Divided: Confronting Inequality in Developing Countries *(New York: 2013).*

As the world's population continues to grow, there is a legitimate concern about depleting Earth's resources faster than they can be replenished. The Global Footprint Network (GFN), an agency that tracks humanity's ecological footprint and nature's capacity to replenish its resources, has found that in just eight months humans exhaust a year's worth of the Earth's bio-capacity.[33] GFN estimates that the world is consuming resources at the rate of 1.5 planets per year.[34] Some studies have argued that the world must replace its growth economy with a steady-state economy, in which production is only replaced, not increased, while the economy continues to develop by improving and renewing its existing resources.[35]

Scientists have observed that the world is entering what they have termed the Anthropocene epoch, to refer to the current state in which humanity is destroying its own habitat. The need to shift away from the consumption-centered paradigm has started to become more appealing in the political spectrum. The United Nations shows that 1 billion out of 7 billion people live below poverty levels, often referred to as the "bottom billion."[36] This group is often the first to experience the negative impacts of the world's rate of development, such as global climate change, water depletion, food shortages, and biodiversity destruction.[37] World population is expected to reach 9.6 billion people by 2050, according to some demographers, with much of that expansion happening in developing countries, whose growth rates will continue to strain Earth's ability to replenish its resources.[38]

Another downside to the current model of measuring well-being is that focusing on growth can lead to greater levels of inequality and social stratification. It can also lead to a tug-of-war between generating output and devoting resources to priorities such as environmental protection, social welfare, and national security.[39] As a result, alternative measures for gauging progress have emerged in recent years. Measures such as the Genuine Progress Indicator account for the social, educational, economic, and environmental activities that contribute to economic growth but that go unnoticed in current national accounting frameworks. For instance, a volunteer's contribution to an organization will not help boost GDP regardless of the fact that this individual puts in the same number of working hours as an employee working the same job in the same organization.[40]

Studies suggest that although people's level of happiness significantly increases when societies develop, high levels of uncertainty and social and economic inequality emerge because of newly developed levels of social status.[41] But a definitive link between increases in a factor such as income inequality and declines in satisfaction has not been fully established.[42] For example, despite the continuous amount of development that has occurred in the United States, France, and Germany over the last two decades, people's happiness levels have seemed to stay the same since 2002.[43] In contrast, Nicaragua, Honduras, El Salvador, and Guatemala rank among the happiest countries in the world despite high levels of poverty and crime.[44]

Commodity Prices Kept Slowing in 2013 But Still Strong Overall

Mark Konold and Jacqueline Espinal

Global commodities markets fell an average of almost 9 percent in 2013, continuing a slowdown that began in 2012.[1] (See Table 1.) The decade prior to 2012 has been called part of a "super-cycle," a 10–35 year trend of rising commodity prices.[2] Previous super-cycles include U.S. economic expansion at the end of the nineteenth century and Europe's post-war reconstruction.[3]

This third cycle has been defined by the surging growth of China since the turn of this century.[4] Since China decided to shift from an export-led growth model, which requires a great deal of natural resources for production, to a model based on internal investment and consumption, the super-cycle has been expected to slow. And it has. But despite this shift as well as geopolitical uncertainty, increasing weather events, and prolonged drought, commodity prices fell marginally in 2013, suggesting that perhaps the super-cycle has been and continues to be driven by other substantial global factors.[5]

Commodities markets are composed of physical goods and raw materials that are bought and sold in large quantities on exchanges around the world. Oil, gold, and agricultural markets traditionally account for the highest volume of trading, but other important assets such as metals, foodstuffs, timber, and fertilizers are also included. (See Figure 1.)

These goods are often traded directly and without a formal exchange market, a process simply known as over the counter (OTC). In 2013, OTC trading volume decreased by 5.3 percent and just over $47 billion was pulled out of the market, blamed partially on a more stable market (lower volatility) and a severely reduced number of investors.[6] By contrast, trading involving commodities derivatives and futures contracts (that is, an attempt to take advantage of the difference between a commodity's current price and its price at some future time) was up 23 percent, led mostly by investors in the United States and China.[7]

It is not clear exactly who benefits from all of this financial activity. Certain instruments like futures contracts in the OTC market were still valued at around $3 trillion in 2013, although that was down from $8 trillion just six years earlier.[8]

Table 1. Change in Commodity Indices Prices, 2003–12 and 2012–13		
	Change 2003–12	Change 2012–13
	(percent)	
Energy	206.1	−0.12
Food	124.6	−7.11
Beverages	100.8	−10.07
Fertilizers	240.2	−17.37
Raw Materials	79.7	−5.85
Precious Metals	376.1	−16.88
Metals	161.0	−5.53
All Commodities	184.1	−8.90

Sources: World Bank, Annual Commodity Indices Price Data 2014.

Mark Konold is Caribbean program manager in the Climate and Energy Program at Worldwatch Institute. **Jacqueline Espinal** is an intern in the Climate and Energy Program.

Persistent global poverty, particularly in countries that provide the raw materials for commodities, would seem to suggest that it is not the source countries that benefit. An investigation of the recent super-cycle in local Brazilian labor markets, however, found that, overall, this rush of economic activity has benefited local laborers in the form of increased wages and a higher presence of "good jobs."[9]

Critical commodities groups—including energy, metals such as copper and gold, and foodstuffs—are closely watched as a bellwether of the overall commodities market. Recent price stability in the oil market continued in 2013 despite uncertainty regarding output from conflict-stricken producers such as Iraq and Libya.[10] (See Figure 2.) After the surge in prices during the first years after 2000, OPEC tried—and through 2013 succeeded—to act as a swing regulator by cutting supply when prices fell and increasing production when prices topped $100–110 per barrel.[11] Tar sands exports in North America and rapidly rising levels of U.S. shale liquids production have also kept oil prices from climbing.[12]

Widespread use of fracking technology, although contentious, has significantly reduced the price of natural gas. The price of natural gas futures on the New York Mercantile Exchange dropped from $15.78 in 2005 to $3.49 per million British thermal units in 2013.[13] Since introducing environmental targets to reduce carbon emissions by 40 percent per unit of gross domestic product from 2005 to 2020, China has been diversifying its energy sources, including increasing the share from natural gas.[14] However, continued economic growth, even if slower than recent years, is expected to result in a doubling of coal consumption.[15] So while China's economic slowdown might have affected global energy markets, it would appear that myriad factors play a role in determining overall prices and quantity levels.

In terms of metals, it is hard to determine if China's economic shift is solely responsible for changes in the global market. Metals continued their recent downward movement, with prices declining 4 percent in 2013.[16] In fact, metals prices have slid 33 percent since 2011.[17] Despite the massive drop, price levels are still considerably higher than they were at the turn of the century.[18] For example, even though the price of copper declined 7.9 percent in 2013, it is still up by more than 300 percent since 2000.[19] (See Figure 3.) Geological factors and higher energy prices, the latter being a main input for metals production, have contributed to

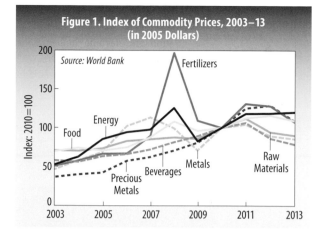

Figure 1. Index of Commodity Prices, 2003–13 (in 2005 Dollars)

Source: World Bank

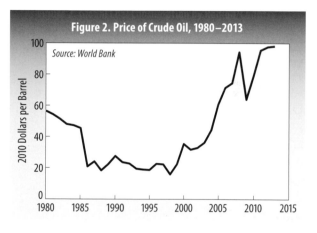

Figure 2. Price of Crude Oil, 1980–2013

Source: World Bank

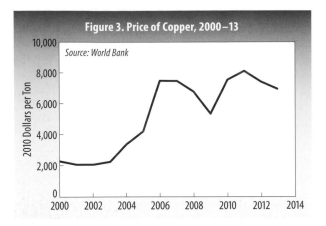

Figure 3. Price of Copper, 2000–13

Source: World Bank

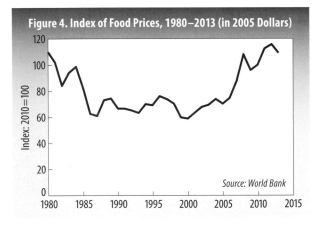

Figure 4. Index of Food Prices, 1980–2013 (in 2005 Dollars)

Source: World Bank

this sustained price level. But so have factors like Indonesia's ban on exports of mineral ores, including nickel, bauxite, and copper.[20] China's economic rebalance was supposed to usher in lower demand for imports, including metals like copper and iron ore. But an additional 260 million rural citizens are expected to relocate to urban areas in China, which will sustain a higher need for such resources.[21]

The big story in the metals markets in 2013 was the precipitous fall in the price of gold. In recent years, gold was used as an investment safe haven against volatile stock markets, but last year saw its biggest annual drop since 1981.[22] This may have had less to do with an economic slowdown in China than with impacts elsewhere. In the United States, the Federal Reserve announced its plan to reduce the rate at which it was buying U.S. bonds, a move that has buoyed the U.S. economy and staved off inflation.[23] In addition, 2013 saw a remarkable surge in equities markets—one in which the percentage gains in stock markets almost exactly matched the percentage drop in gold, 30 percent.[24]

Despite a 7.1-percent drop in 2013, food prices remain at historic highs and are still up an average of 9.2 percent a year (5.8 percent a year in real terms) since the turn of the century.[25] (See Figure 4.) In China, environmental strain has led to increased imports of wheat, corn, and rice to support a growing population.[26] This import dependence affects domestic and international food market prices, as the country continues to be a high importer of agricultural goods.[27] And though this does affect global prices, it would appear wider-ranging factors are influencing overall prices: high demand for livestock feed, renewable fuel standards (supporting fuels that rely heavily on agricultural products as inputs), declines in global buffer stocks, and policy choices such as export bans in some regions are all working to keep food prices high.[28]

Further, increasing extreme weather events due to global climate change have negatively affected agricultural yields, which only exacerbates the problem.[29] Expected biofuel production, hydroelectric dam construction, and further urbanization in developing countries are likely to reduce available cropland by, respectively, 15 million, 10 million, and 30 million hectares by 2030.[30]

It is still too early to tell if China's economic shift is the most decisive factor in the direction of global commodities. Clearly, the choice to become a more consumer-driven rather than an export-driven economy has affected areas such as energy, metals, and food. However, data from all countries continue to show that perhaps more than just one factor creates a "super-cycle," as price levels remain significantly higher than they have been in decades past.

Paper Production Levels Off

Michael Renner

In wealthy countries, paper is so ubiquitous that it comes in seemingly endless numbers of grades, types, shapes, and colors and is used for both mundane and highly specialized purposes. Too often, paper products are discarded soon after their purchase, and only a portion is recovered for recycling.

According to the United Nations Food and Agriculture Organization (FAO), 397.6 million tons of paper and paperboard were produced worldwide in 2013, the latest year for which global data are available.[1] Production thus declined slightly for the second year in a row, from an all-time high of 400.6 million tons in 2011.[2] Since the global economic crisis of 2008, paper production has leveled off, following a more than fivefold rise over the past half-century.[3] (See Figure 1.)

Most of the paper produced today is used for wrapping and packaging purposes, with its share rising from 39 percent of total output in 1961 to 54 percent in 2013.[4] The share of printing and writing grades peaked at 30 percent by 2000 and has since declined to 26 percent.[5] Newsprint once accounted for a quarter of all paper production, but now this is down to just 7 percent.[6] Production of household and sanitary tissues has expanded from about 4 percent to almost 8 percent.[7]

Just four countries—China, the United States, Japan, and Germany—together account for more than half (56 percent) of the world's paper production.[8] (See Figure 2.) The United States was historically by far the largest producer, with as much as 42 percent of the global total in the early 1960s.[9] But its output peaked in 1997 at 88.5 million tons.[10] The United States and Japan have both been eclipsed by China.[11]

China's production expanded rapidly after 2000. Supported by at least $33

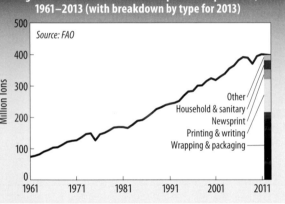

Figure 1. Global Production of Paper and Paperboard, 1961–2013 (with breakdown by type for 2013)

Source: FAO

Other
Household & sanitary
Newsprint
Printing & writing
Wrapping & packaging

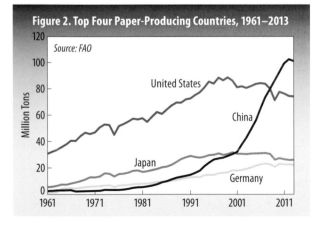

Figure 2. Top Four Paper-Producing Countries, 1961–2013

Source: FAO

United States
China
Japan
Germany

Michael Renner is a senior researcher at Worldwatch Institute and codirector of *State of the World 2015*.

billion in government loans and subsidies, output more than tripled to 102.5 million tons in
2012.[12] A slight dip in 2013, to 101.1 million
tons, was only the second time that production declined in China in the last half-century.[13]
Accompanying the growth in Chinese output
was a shift from non-wood fibers such as cereal
straws and bamboo to wood fibers, the development of large-scale tree plantations, and the
consolidation of a once-fragmented industry.[14]

For many years Canada was the second-
or third-largest producer.[15] However, it was
bypassed not only by China but also by Germany and South Korea.[16] After 2000, Canada's
output plummeted by almost half, from 20.9
million tons to 11.3 million tons.[17] Close
behind are Sweden and Finland, as well as
Brazil, India, and Indonesia—three countries
whose production has risen strongly.[18] The
top 11 producers together accounted for three-
quarters of global production. (See Figure 3.)
South American producers are adding to their
production capacity, based on fast-growing
eucalyptus and pine monoculture plantations.[19]

Global exports of paper and paperboard
climbed more than ninefold between 1961 and
2007 to a volume of 117.6 million tons, but
trade flows have since been wobbly, declining
to 109.4 million tons in 2013.[20] Relative to production, exports almost doubled from about 17
percent in the early 1960s to 31.2 percent in

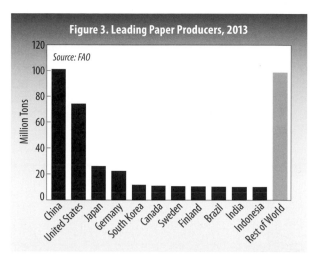

Figure 3. Leading Paper Producers, 2013

Source: FAO

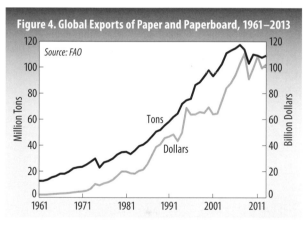

Figure 4. Global Exports of Paper and Paperboard, 1961–2013

Source: FAO

2004, then dropped to 27–28 percent.[21] By value, paper exports followed a less
steady trajectory, but they rose from about $2.1 billion in 1961 to $102.0 billion in
2013, with a peak at $110.8 billion in 2008.[22] (See Figure 4.)

Exports are dominated by Europe, which accounted for 60 percent of global
shipments in 2013, followed by North America with 19 percent and Asia with 17
percent.[23] But Europe also leads in imports (51 percent), ahead of Asia's 23 percent
and North America's 12 percent.[24] Five countries—Germany, the United States,
China, France, and Italy—are both top exporters and top importers.[25]

The top 100 companies produced 212.7 million tons of paper in 2013, or
53 percent of the world total, and commanded sales of more than $320 billion.[26]
These companies are headquartered in 26 countries.[27] (See Table 1.)

The top 10 firms hail from the United States, Finland, Japan, Ireland, Sweden,
South Africa, the United Kingdom, and China.[28] (See Table 2.) Table 2 is somewhat incomplete, however, since relevant data are not available for some leading companies. For example, an annual listing by *PPI Magazine* has sales but not

Table 1. Regional Distribution of Top 100 Paper-Producing Companies, and Sales and Production, 2013

Region	Number of Companies	Sales	Production
		(billion dollars)	(million tons)
Europe	33	108.2	70.6
North America	29	116.9	62.0
Asia	27	69.1	67.8
Latin America	8	17.3	5.7
Africa	2	6.5	6.7
Oceania	1	2.9	0*
Total	100	320.9	212.7

* Production of wood pulp only.
Totals may not add due to rounding.
Source: Graeme Rodden, Mark Rushton, and Annie Zhu, "PPI Top 100: A Newcomer from Brazil is Welcomed to the List Along with Two from China," PPI Magazine, 1 September 2014.

Table 2. Top 10 Paper Companies by Sales and Production, 2013

Company	Headquarters	Sales		Production	
		(billion dollars)	(top 10 rank)	(million tons)	(top 10 rank)
International Paper	United States	29.1	1	19.6	1
Procter & Gamble	United States	16.8	2	na	na
UPM	Finland	13.1	3	10.3	3
Stora Enso	Finland	12.8	4	9.9	4
Oji Paper	Japan	11.6	5	8.7	5
Smurfit Kappa	Ireland	10.6	6	7.0	7
Kimberly Clark	United States	10.0	7	na	na
Marubeni	Japan	9.8	8	0.6	-
SCA	Sweden	9.6	9	5.1	10
Rock Tenn	United States	9.1	10	8.1	6
Nippon Paper	Japan	8.7		6.9	8
Mondi	U.K./S. Africa	8.6		5.3	-
Sappi	South Africa	5.9		6.7	9
Nine Dragons	China	4.6		11.1	2

Note: Companies in italics are among top 10 in terms of production but not sales.
Source: Compiled from Graeme Rodden, Mark Rushton, and Annie Zhu, "PPI Top 100: A Newcomer from Brazil is Welcomed to the List along with two from China," PPI Magazine, 1 September 2014.

production data for such large companies as U.S.-based Procter & Gamble and Kimberly Clark. A similar annual survey by PricewaterhouseCoopers also does not have data for a number of companies that would warrant inclusion, such as Asia Pulp and Paper (from Indonesia/China).[29]

Paper consumption in North America, Europe, and Japan has declined in recent years, shifting to other parts of the world. The share used in industrial countries fell from 66 percent of global demand in 1995 to 42 percent in 2013.[30] China's share rose from 9 percent to 25 percent during the same period; in 2012, all of Asia accounted for 46.1 percent, compared with 24.2 percent in Europe and 19.4 percent in North America, while the rest of world came to only 10.4 percent.[31]

But on a per capita basis, wealthy countries continue to use far greater amounts than developing countries do—on average, 221 kilograms (kg) in 2012 in North America and 125 kg in Europe compared with 45 kg in Asia, 43 kg in Latin America and the Caribbean, and just 7 kg in Africa.[32] (See Figure 5.)

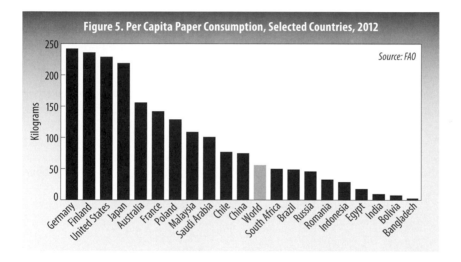

Figure 5. Per Capita Paper Consumption, Selected Countries, 2012

Source: FAO

A considerable share of paper usage in rich countries is short-lived and wasteful, whether in the form of unnecessary packaging, printing of e-mails and other materials even in an age of electronic communications, or unwanted commercial mailings. For instance, Americans receive an estimated 85 billion pieces of junk mail, weighing about 4.7 million tons, each year.[33]

Paper recycling helps to reduce energy use and pollution. On average fibers can be recycled five to seven times before they become unusable. Recovered paper with long fibers (such as office printing paper) is generally of higher value than paper with short fibers (such as newsprint).[34] Recycled paper requires 60 percent less energy and 80 percent less water to produce than virgin paper, and it generates 95 percent less air pollution.[35] Recycling one ton of paper on average saves 26,500 liters of water, about 318 liters of oil, and 4,100 kilowatt-hours of electricity.[36]

According to FAO, recovered paper production (which includes residues from production processes and post-consumer recycling) was 215 million tons in 2013, equal to 54 percent of total world paper supply.[37] This is up from about 20 percent in the early 1960s.[38] Paper produced from virgin wood pulp supplied 174 million tons, and non-wood pulp accounted for another 14 million tons.[39]

The leading producers of recovered paper in 2013 were China and the United States (with 23 and 21 percent of the global total, respectively).[40] They were followed by Japan (10 percent), Germany (7 percent), South Korea (4 percent), and the United Kingdom (4 percent).[41] About a quarter of recovered paper is traded internationally (shipped mostly to Asia), up from just 6 percent in 1970.[42]

Paper recovery plays a vastly different role in individual producer countries. Japan got close to 84 percent of its paper production from recovered paper in 2013, Germany and South Korea had 69 and 68 percent, respectively, and the United States 61 percent.[43] They are followed by China (45 percent), Brazil (43 percent), and Indonesia (38 percent).[44] In other countries, the share is much lower—26 and 24 percent, respectively, in India and Canada, and a diminutive 11 and 6 percent, respectively, in Sweden and Finland.[45]

The pulp and paper industry is a large energy and water consumer, as well as a user of toxic chemicals. In the United States, the paper industry is the third-largest energy user among manufacturing industries, accounting for 11 percent of energy consumption in 2010, the most recent year for which such information is available.[46]

Chlorine bleaching of paper can result in the formation of dioxins and furans.[47] Since the 1990s, most mills in Europe and North America, as well the modern ones in China, have moved either toward elemental chlorine—a safer, though still harmful, process—or toward totally chlorine-free paper, which is either unbleached or bleached with oxygen, ozone, and/or hydrogen peroxide.[48] In 2005, elemental chlorine was used in about 20 percent of kraft pulp production globally, down from more than 90 percent in 1990.[49]

Chinese paper producers have lagged in efficiency. In 2010, a typical U.S. or European paper mill used 0.9–1.2 tons of coal and about 35–50 tons of water per ton of pulp, while Chinese mills averaged 1.4 tons of coal and 103 tons of water.[50] However, to reduce emissions and effluents, the Chinese government mandated the closure of the most polluting mills, those producing 25 million tons of pulp and paper, between 2007 and 2012 alone, while more modern mills opened up.[51]

Reducing the paper industry's environmental footprint requires continued progress in minimizing paper consumption and avoiding waste, in raising paper recovery and recycled content, in ensuring that virgin fiber is derived from sustainable forestry and logging operations, and in using less polluting and less energy-intensive paper production methods.[52] Some progress on these fronts has been made, though mostly at a slow pace.

Global Plastics Production Rises, Recycling Lags

Gaelle Gourmelon

For more than 50 years, global production of plastics has continued to rise.[1] Some 299 million tons of plastics were produced in 2013, representing a 3.9-percent increase over 2012's output.[2] With a market driven by consumerism and convenience, along with the comparatively low price of plastic materials, demand for plastic is growing.[3] Recovery and recycling, however, remain insufficient, and millions of tons of plastics end up in landfills and oceans each year.[4]

Plastics are human-made materials manufactured from polymers, or long chains of repeating molecules.[5] They are derived from oil, natural gas, and—while still a small portion of overall production—increasingly, from plants like corn and sugarcane.[6] About 4 percent of the world's petroleum is used to make plastic, and another 4 percent is used to power plastic-manufacturing processes.[7]

First invented in the 1860s but developed for industry in the 1920s, plastics production exploded in the 1940s, becoming one of the fastest-growing global industries.[8] From 1950 to 2012, plastics growth averaged 8.7 percent per year, booming from 1.7 million tons to the nearly 300 million tons of today.[9] Worldwide production continued to grow between the 1970s and 2012 as plastics gradually replaced materials like glass and metal.[10] Metal, glass, and paper are increasingly replaced by plastic packaging, particularly for foods. By 2009, plastic packaging accounted for 30 percent of packaging sales.[11] With the push by U.S. federal mileage standards to reduce the weight of vehicles, the American automobile industry has been a champion of this transition, too. Plastics make up about 10 percent by weight (50 percent by volume) of a typical U.S. vehicle today, representing 336 pounds of plastic per vehicle.[12] In 1960, less than 20 pounds of plastic were used in cars.[13] Plastics now often replace metals in bumpers and door panels as well as in engine components.

Today, the global plastics industry generates revenue of about $600 billion annually.[14] Plastics are found throughout many sectors and industries, including transportation, construction, health care, food products, telecommunications, and consumer goods.[15] Per capita, plastics consumption reached 100 kilograms (kg) in Western Europe and North America.[16] Asia currently uses just 20 kg per person, but this figure is expected to grow rapidly.[17]

Plastics production is also shifting toward Asia. The region produced 45.6 percent of global plastics in 2013, with China alone producing nearly a quarter of the world's plastics.[18] (See Figure 1.) China surpassed Europe in plastics production in 2010.[19] India has recently seen strong growth in plastics production due to an increasing population and the growth of manufacturing sectors in the country.[20] Today, Europe and the states emerging from the former Soviet Union account for

Gaelle Gourmelon is the communications and marketing manager at Worldwatch Institute.

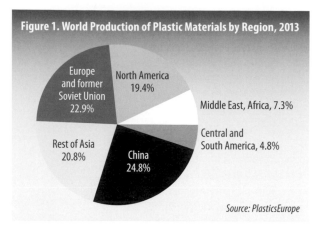

Figure 1. World Production of Plastic Materials by Region, 2013

Europe and former Soviet Union 22.9%

North America 19.4%

Middle East, Africa, 7.3%

Central and South America, 4.8%

Rest of Asia 20.8%

China 24.8%

Source: PlasticsEurope

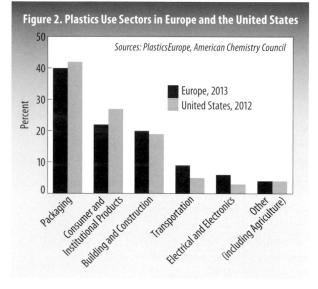

Figure 2. Plastics Use Sectors in Europe and the United States

Sources: PlasticsEurope, American Chemistry Council

Europe, 2013
United States, 2012

Percent

Packaging
Consumer and Institutional Products
Building and Construction
Transportation
Electrical and Electronics
Other (including Agriculture)

22.9 percent of global plastics production, with Germany leading European production. North America, led in large part by the United States, produces 19.4 percent of global plastic.[21] The Middle East and Africa (7.3 percent) and Central and South America (4.8 percent) have the smallest global shares of production.[22]

Packaging is responsible for the majority of plastics use, representing 40 percent of demand in Europe and 42 percent in the United States.[23] (See Figure 2.) Consumer and household products (like appliances, toys, plastic cutlery, and furniture) account for the next most significant segment, closely followed by the building and construction sector.[24] Within the United States, the largest consuming industries are construction (which uses plastic products like pipes, window and door frames, roofing, and siding) and the food and beverage industry (which uses products like plastic bottles, containers, and film).[25]

Between 22 and 43 percent of plastic worldwide is disposed of in landfills, where its resources are wasted, it takes up valuable space, and it blights communities.[26] Recovering plastic from the waste stream for recycling or for energy has the potential to minimize these problems.[27] However, much of the plastic collected for recycling is shipped to countries with lower environmental standards, making the balance between environmental protection, clean material cycles, and resource use unstable. Further, energy recovery from plastic is often inefficient, requires air emissions controls, and produces hazardous ash.[28] Actual rates of recovery vary widely across countries.

In Europe, 26 percent, or 6.6 million tons, of post-consumer plastic produced in 2012 was recycled, while 36 percent was incinerated for energy recovery.[29] This means that 38 percent of post-consumer plastics in Europe went to landfills.[30] This represents a 26-percent decrease in plastic ending up this way compared with 2006, but nearly half of all European countries still send the majority of their plastics waste to landfills.[31] Nine European countries—Austria, Belgium, Denmark, Germany, Luxembourg, the Netherlands, Norway, Sweden, and Switzerland—have enacted landfill bans for plastics.[32] While these countries do generally achieve higher recycling rates than countries with no landfill bans, a majority of plastic is disposed of through incineration for energy recovery.[33]

In the United States, only 9 percent of plastic (2.8 million tons) was recycled in 2012.[34] The remaining 32 million tons were discarded, accounting for nearly 13 percent of the nation's municipal solid waste stream.[35] The United States depends mostly on China and Hong Kong to absorb its plastic waste, although some is sent to Canada and Mexico.[36]

In other parts of the world, recovery of plastics is even lower. The United Nations Environment Programme estimates that 57 percent of plastic in Africa, 40 percent in Asia, and 32 percent in Latin America is not even collected, being instead littered or burned in the open.[37]

The largest waste plastics exporters are the United States, followed by Japan, Germany, and the United Kingdom.[38] Europe exports about half of the plastic it collects for recycling and is the largest global exporter of waste plastic intended for recycling.[39]

Most plastic scraps from western countries with established collection systems flow to China, which receives 56 percent (by weight) of waste plastics imports worldwide.[40] What happens to all this imported plastic once it gets there is not well understood, however.[41] The International Solid Waste Association reports that indirect evidence suggests the majority of plastic is still being reprocessed using family-run, low-tech businesses with no environmental protection controls.[42] There are also concerns that low-quality plastic is not reused but is disposed of or incinerated for energy recovery in plants without air pollution control systems.[43] Through its 2010 Green Fence Operation, the Chinese government has started to work to reduce unregulated facilities.[44]

Europe depends on China for 87 percent of its plastic waste exports intended for recycling.[45] This leaves the global plastics recycling market highly vulnerable to market changes in China, as was shown by recent quality control implementations by China's customs (implementing a "zero tolerance" policy for contamination levels in imports), which resulted in a crash in the price of secondary raw materials.[46]

Approximately 10–20 million tons of plastic end up in the oceans each year.[47] Including financial losses by fisheries and tourism as well as time spent cleaning beaches, $13 billion a year is lost in environmental damage by plastics to marine ecosystems.[48] Marine wildlife is particularly vulnerable to plastics pollution.[49] Animals such as seabirds, whales, and dolphins can become entangled in plastic matter.[50] Floating plastics—like discarded nets, docks, and boats—transport microbes, algae, invertebrates, and fish into non-native regions.[51]

Once in the ocean, plastic does not go away. It breaks down into small pieces that are ingested by sea life and transferred up the food chain, carrying chemical pollutants from prey to predator.[52] A recent study conservatively estimated that 5.25 trillion plastic particles weighing 268,940 tons are currently floating in the world's oceans.[53] Because plastics are moved with wind and currents, very few areas in the ocean may have escaped plastics pollution.[54] The North Pacific gyre contains the most, with nearly 2 trillion pieces of plastic weighing over 96,000 tons.[55] (See Figure 3.) In the southern hemisphere, the Indian Ocean has more plastics pollution than the South Pacific and South Atlantic combined do.[56]

The environmental and social benefits of plastic must be weighed against the problems that its durability and incredible volume worldwide constitute as a waste stream.[57] Plastics help to reduce food waste by keeping products fresh longer, allow

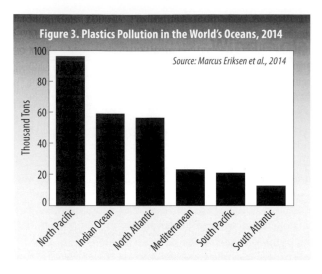

Figure 3. Plastics Pollution in the World's Oceans, 2014

Source: Marcus Eriksen et al., 2014

for the manufacture of health care equipment, reduce packaging mass compared with other materials, improve transportation efficiency, and have a large potential for renewable energy technologies.[58] But plastics litter, gyres of plastics in the oceans, and toxic additives in plastic products—such as colorings, flame retardants, and plasticizers—are raising awareness and strengthening consumer demand for more-sustainable materials.[59]

Along with reducing unnecessary plastics consumption, finding more environmentally friendly packaging alternatives, and improving product and packaging design to use less plastic, many challenges associated with plastic could be addressed by improving management of the material across its lifecycle.[60] Businesses and consumers could increase their participation in collection in order to move plastic waste toward a recycling or recovery supply chain. Companies could switch to recycled plastics, using joint ventures to ensure supply.[61] They could also investigate options for using bioplastics—plastics that are partly or fully biobased, biodegradable, or both—although the benefits and impacts of these products are currently unknown.[62] Governments must regulate the plastics supply chain to encourage recycling, and consortia must coordinate and monitor the supply chain and provide guidelines for plastics waste processing, especially in developing economies.[63] As the economy and population grow, global demand for plastics is expected to continue to grow—especially in Africa, Latin America, the Middle East, and China.[64] Immediate action is needed to handle today's and tomorrow's plastics problem.

Population and Society Trends

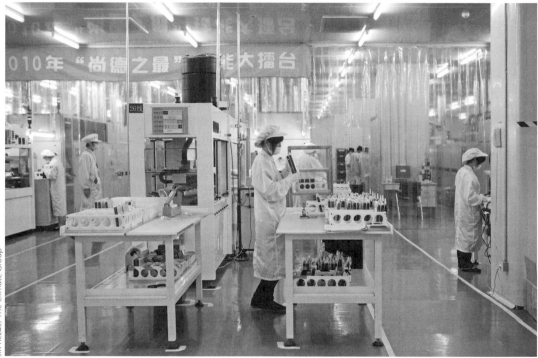

Solar cell manufacturing lines at the Suntech Solar Energy factory in Wuxi, China

For additional population and society trends, go to vitalsigns.worldwatch.org.

Will Population Growth End in This Century?

Robert Engelman and Yeneneh Terefe

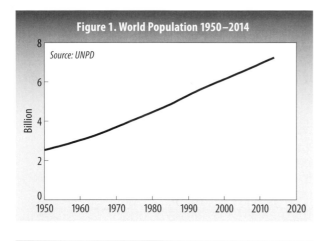

Figure 1. World Population 1950–2014

Source: UNPD

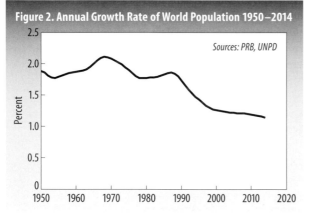

Figure 2. Annual Growth Rate of World Population 1950–2014

Sources: PRB, UNPD

Having nearly tripled from 2.5 billion people in 1950 to 7.3 billion today (see Figure 1), human population will continue growing through 2070, according to two recent demographic projections.[1] After that, population either will begin to shrink or will continue growing into the next century, depending on which of the two projections more accurately forecasts the future.

In the years following World War II, population grew fairly rapidly, with a rate of growth that peaked in the late 1960s at 2.1 percent a year.[2] (See Figure 2.) Since then population growth has gradually slowed—although with a larger base each year, the number of people added annually has changed little. Every year sees the addition of about 80 million human beings on the planet, roughly the current population of Germany, Turkey, or Egypt.[3]

The two population projections—one from the United Nations Population Division (UNPD), the other from the International Institute for Applied Systems Analysis (IIASA)—agree on how population has grown until now.[4] But their future scenarios document a breakdown in consensus among demographers about the future. The people-counting social science seems to be entering a new realm in which scientists recognize how much uncertainty the world and its population faced in 2014. Unusually for the demographic discipline, experts even went public with their disagreement about the most likely trends in the critically influential area of how many people will live on the planet in the future: letters to the editor of both the *Wall Street Journal* and *Science* by key authors of both sets of projections laid out some of the reasoning behind their numbers.[5]

Working with colleagues at the University of Washington, U.N. demographers in 2014 for the first time used probabilistic projections, a methodology that applies past behavior and expert opinion about the future to assign quantified

Robert Engelman is a senior fellow and former president of Worldwatch Institute. **Yeneneh Terefe** is a research assistant at the Institute.

probabilities to various population outcomes. Based on this, in September 2014, they claimed there was an 80-percent likelihood that world population would grow from today's 7.3 billion people to 10.9 billion people in 2100, plus or minus about 1.3 billion.[6] Defying a widespread news media and public perception that a stationary world population of 9 billion in 2050 was a near certainty, the U.N. analysts reported that the most likely long-term future is for continued growth into the twenty-second century.[7]

Demographers associated with IIASA, based in Laxenburg, Austria, begged to differ with this analysis, however. They have long used the probabilistic methodology, and in October 2014 they released their most recent set of population projections.[8] These foresee world population peaking around 2070 at 9.4 billion people and then gradually shrinking to 8.9 billion by the century's end. (See Figure 3 for the two contrasting population curves.)[9]

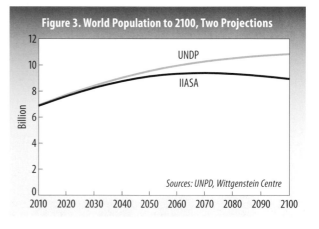

Figure 3. World Population to 2100, Two Projections

Sources: UNPD, Wittgenstein Centre

What explains the disagreement between these two respected groups of population projectors? In two words: different assumptions, revolving mostly around two topics—Africa and the future of education. The U.N. demographers stressed that recent surveys have shown that human fertility (defined demographically as the average number of children that women in a population give birth to over their lifetimes) is not falling in a number of countries as earlier projections had assumed they would.[10] Most of these countries are in Africa, and UNPD significantly increased its earlier projections for that continent's population. The new probabilistic projections gave an 80-percent chance that African population will rise from 1.2 billion today to somewhere between 3.5 billion and 5.1 billion by 2100, with a central projection of 4.2 billion.[11] The prospect of a tripling or more of Africa's population during this century generated considerable news media attention when the U.N. numbers were released.[12]

The IIASA demographers, by contrast, focused not on recent fertility trends so much as educational ones. In every region of the world, including Africa, the proportions of young people enrolled in school have generally been rising. The IIASA analysts documented rising rates of school attainment—the highest average level of schooling attained by populations—and argued that these rates are likely to continue to rise.[13] Since even moderately high levels of educational attainment are associated with reductions in fertility, the demographers reasoned, fertility even in high-fertility countries is likely to fall more than current fertility trends suggest on their own. This different focus led in particular to especially divergent projected paths for population in Africa, with the Austrian demographers projecting a population of 2.6 billion by 2100.[14] (See Figure 4.)

The differences in the two groups' projections for other world regions are less dramatic, since fertility has already fallen considerably in recent decades in the rest

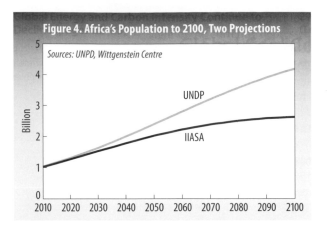

Figure 4. Africa's Population to 2100, Two Projections

Sources: UNPD, Wittgenstein Centre

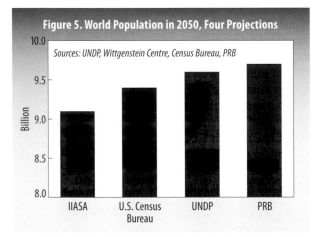

Figure 5. World Population in 2050, Four Projections

Sources: UNDP, Wittgenstein Centre, Census Bureau, PRB

of the world, education levels are already fairly high on average, and there is thus less division on where population is likely to go. This is evident in the two groups' projections for Asia, which has more than half of the world's population today but is projected by both sets of demographers to peak in population in the 2050s, followed by a gradual decline.[15]

The two groups' medium or "best-guess" projection for the world's major regions besides Africa are as follows: For Asia, the U.N. Population Division projects 4.7 billion people in 2100, while IIASA projects 4.4 billion.[16] For Europe, UNPD projects 639 million and IIASA, 702 million.[17] For North America, UNPD has 513 million, while IIASA projects 520 million.[18] And for South America and the Caribbean, UNPD 736 million, IIASA 684 million.[19]

Two other organizations, the U.S. Census Bureau and the Population Reference Bureau (PRB), also project world population, though in both cases only up to 2050.[20] Some demographers believe that there is too much uncertainty about demographic trends and the conditions that influence them to go beyond the next three or four decades in projections. Even so, lining up all four of these projections for 2050 provides yet another reminder of modest differences in demographers' best guesses about the future of population, even for a point just 35 years away, ranging from IIASA's 9.1 billion to PRB's 9.7 billion.[21] (See Figure 5.) The difference between these two "best guesses" of mid-century population, at more than 540 million people, is considerably greater than the current population of the United States (323 million).[22]

Two Australian environmental scientists recently published a fifth set of population projections—with a twist.[23] Demographers' projection scenarios usually vary by fertility or migration rates but assume continually falling death rates. The reasoning is that there is no way to assess the likelihood of increases in mortality, given the long historic trend of rising life expectancies for the human species as a whole and in almost all countries. Scientists Corey J. A. Bradshaw and Barry W. Brook went against this longstanding practice. They based their population projections on those of UNPD but added scenarios in which humanity experienced increases in the deaths of children due to climate change or outright demographic catastrophes due to "global pandemic or war." In their most extreme scenario, 6 billion people die in the early 2040s.[24] With that as a backdrop assumption and with fertility trends based on a standard middle scenario, the two scientists projected that human population would not recover its 2014 population in this century but rather reach 2100 with about 5 billion people.[25]

Even leaving aside this projection outlier (the Australian analysts are non-demographers engaging in a one-off thought exercise), the significant differences among demographers' projections tell us something important about population and the human future. Despite general perceptions that demographers confidently forecast future population, no one knows when population will stop growing or the level it will peak at. A further implication is that the future of population growth may respond to decisions made today. Ideally, these will be decisions that support a reduced incidence of unintended pregnancy (now about 40 percent of all pregnancies globally) rather than ones that allow environmental and social conditions to deteriorate until death rates reverse their historic decline.[26]

Jobs in Renewable Energy Expand in Turbulent Process

Michael Renner, Rabia Ferroukhi, Arslan Khalid, and Alvaro Lopez-Peña

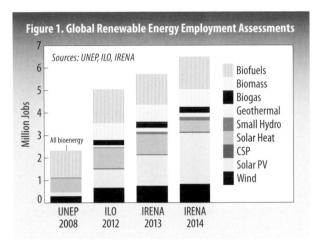

Figure 1. Global Renewable Energy Employment Assessments

Sources: UNEP, ILO, IRENA

Legend: Biofuels, Biomass, Biogas, Geothermal, Small Hydro, Solar Heat, CSP, Solar PV, Wind

This article was prepared in collaboration with IRENA. It draws on research carried out for IRENA's recent publication, *Renewable Energy and Jobs: Annual Review 2014*. **Michael Renner** is a senior researcher at Worldwatch Institute. As a consultant to IRENA, UNEP, and ILO, he coauthored the global renewable energy employment reports cited in this article. **Rabia Ferroukhi** is deputy director of IRENA's Knowledge, Policy and Finance Division in Abu Dhabi. **Arslan Khalid** is an associate program officer and **Alvaro Lopez-Peña** is a program officer at IRENA.

According to the International Renewable Energy Agency (IRENA), there may now be as many as 6.5 million direct and indirect jobs in renewable energy.[1] This figure is an update of IRENA's previous estimate of 5.7 million such jobs, which was published in December 2013.[2] Earlier assessments by the United Nations Environment Programme (UNEP) in 2008 and by the International Labour Organization (ILO) in 2012 had put the global estimate at 2.3 million and 5 million, respectively.[3] (See Figure 1.)

Although these estimates suggest a strong employment expansion, the figures are not strictly year-by-year comparisons but rather successive efforts to broaden the coverage of countries and to improve and refine data. Although they are based on an extensive data collection, data gaps remain and the underlying studies used in compiling the global estimates are inevitably of uneven detail and quality.[4]

The overall upward trend has been accompanied by considerable turmoil in some renewable energy industries—a reflection of fluctuations in costs, prices, and investments; increasing labor productivity; regional shifts; and waves of industry consolidation and realignment.[5] Austerity measures in the wake of the world economic crisis have also played a critical role.[6] Furthermore, uncertainties or frequent changes in government policies do not favor job creation.[7]

Nowhere are the upheavals more noticeable than in the solar photovoltaic (PV) sector, where intensified competition, massive overcapacities, and tumbling prices have caused a high degree of turbulence in the last two to three years, but have also triggered a boom in installations.[8] In 2013, China became the world's leading installer of solar panels.[9] Global PV employment is thought to have expanded from 1.4 million jobs in 2012 to perhaps as many as 2.3 million in 2013.[10]

Solar PV has thus bypassed biofuels (ethanol and biodiesel) as the top job generator. Most of the 1.45 million biofuels jobs are found in the growing and harvesting of feedstock such as sugarcane, corn, or palm oil.[11] This involves physically demanding manual work, and workers often contend with oppressive workplace conditions.[12] Processing of the feedstock into fuels offers far fewer jobs, but the ones created are higher-skilled and they pay better.[13]

Employment in the next-largest renewables sector, wind power, is estimated to run to some 834,000 jobs.[14] Uncertainty about the future direction of policies in several countries weakened job creation in this field in 2013, leading to a sharp drop in new installations in the United States and to weak markets in large parts of Europe and in India.[15] In contrast, developments in China and Canada were more positive.[16] Europe remains the leader in developing the still-small offshore wind sector, with an estimated 58,000 jobs.[17]

Employment information for the other renewable energy technologies is far more limited. A rough estimate for the biomass industry (for power and heat) suggests that employment may run to about 780,000 jobs worldwide.[18] For the solar heating/cooling industry, the International Energy Agency offers a global estimate of 420,000 jobs.[19] On the basis of recent Chinese estimates, IRENA puts the global total somewhat higher, at about 500,000 jobs.[20] The remaining renewable energy technologies—biogas, small hydropower, geothermal energy (power and heat), and concentrated solar power (CSP)—are much smaller employers.[21]

A small number of countries—China, members of the European Union (EU), Brazil, the United States, and India—account for the bulk of renewable

Table 1. Renewable Energy Jobs, World and Leading Countries, 2012–13						
	World	China	European Union	Brazil	United States	India
			(thousand)			
Biomass*†	782	240	311		152	58
Biofuels	1,453	24	108	820	236	35
Biogas	264	90	69			85
Geothermal*	184		97		35	
Small Hydropower‡	156		26	12	8	12
Solar PV	2,273	1,580	252		143§	112
CSP	43		28			
Solar Heating/ Cooling	503	350	43	30**		41
Wind Power	834	356	308	32	51	48
Total	6,492	2,640	1,242	894	625	391

* Power and heat applications.   † Traditional biomass is not included.   ‡ Large hydro imposes severe environmental and social impacts; although a capacity of 10 megawatts is often used as a threshold separating small from large hydro, there is no universally accepted definition.
§ All solar technologies combined.   ** Equipment manufacturing only; no data available for installation jobs.
Notes: Data are principally for 2012–13, with dates varying by country and technology. Totals may not add due to rounding.
Source: Adapted from IRENA, Renewable Energy and Jobs: Annual Review 2014 (Abu Dhabi: 2014).

energy employment: 5.8 million direct and indirect (supply chain) jobs out of 6.5 million worldwide.[22] (See Table 1.) These countries are home to half of the world's population.[23]

Due to its dominance of the solar PV and solar heating/cooling sectors and its strong presence in wind energy, China is the largest employer in the renewable energy sector. The latest estimates by the country's National Renewable Energy Center suggest a significant increase from earlier PV employment estimates, to almost 1.6 million jobs in PVs in 2013.[24] Other major sources of employment are the wind, solar water heating/cooling, and biomass industries, together providing close to 1 million jobs.[25]

EU member states had more than 1.2 million renewable energy jobs in 2012.[26] The largest employers are the wind, solar PV, and biomass sectors.[27] Wind and biomass posted significant job gains, although the wind industry's prospects now depend significantly on developments in just two countries—Germany and the United Kingdom.[28] In sharp contrast, the solar PV industry in the EU experienced large job losses: substantial reductions in Germany, France, Italy, and some other countries could not be offset by smaller gains in Bulgaria, Denmark, the Netherlands, and Slovenia.[29]

Even though it suffered some job losses in 2013, Germany remains the dominant force in Europe, with about 371,000 direct and indirect jobs.[30] (See Figure 2.) Whereas wind reached a new high of 138,000 jobs in 2013, solar PV employment in Germany plummeted to just 56,000 from a peak of 111,000 jobs in 2011.[31] PV manufacturing was affected even more severely. In a single year, from late 2012 to late 2013, the number of jobs dwindled from more than 12,000 to less than 5,000.[32]

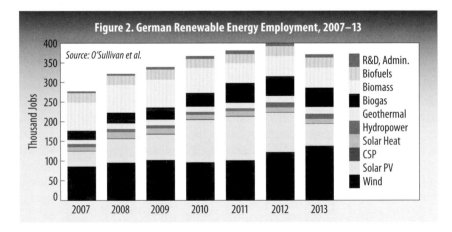

Figure 2. German Renewable Energy Employment, 2007–13

Spain's renewable energy sector has been hit hard by economic crisis and a series of adverse government policy changes.[33] The country suffered a net loss of 23,700 jobs between 2008 and 2012, or 17 percent.[34] (See Figure 3.) The largest losses occurred in the wind power sector (18,000 jobs) and in PV (16,500 jobs).[35] CSP continued to grow until 2011, but then lost close to 6,000 jobs in 2012.[36]

In Brazil, renewable energy is largely synonymous with sugarcane-based ethanol. A factor of rising importance is the growing mechanization of sugarcane harvesting. It has brought the number of direct jobs down from 460,000 in 2006 to 331,000 in 2012, even as ethanol processing jobs increased from 177,000 to 208,000.[37] (See Figure 4.) At about 82,000 jobs, biodiesel production still offers far less employment.[38] Nevertheless, this figure represents a 3.5-fold increase from 24,660 in 2008, and employment in this area might grow to some 460,000 jobs by 2020.[39] Wind power employment is much smaller, but it has jumped from about 3,700 jobs in 2010 to 32,000 jobs in 2013.[40]

In the United States, the number of wind and ethanol jobs has fluctuated, but solar employment has been rising fast.[41] (See Figure 5.) The latest available survey puts solar employment at close to 143,000 jobs in 2013, a gain of 20 percent.[42] Almost half of these are installation jobs; sales and project development provide another 22 percent, while manufacturing is down to only 21 percent from 36 percent in 2011 (reflecting a drop of about 8,000 jobs).[43] In the wind sector, the stop-and-go nature of the U.S. Production Tax Credit has resulted in considerable job fluctuations. A 92-percent decline in new wind installations during 2013 meant that employment fell from 80,700 to 50,500 jobs.[44] U.S. ethanol employment had declined from 181,300 in 2011 to 173,700 during 2012—owing to

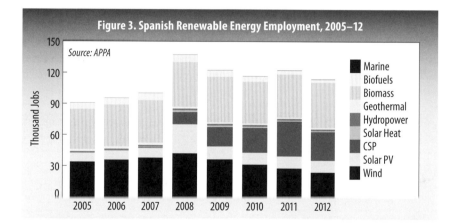

Figure 3. Spanish Renewable Energy Employment, 2005–12

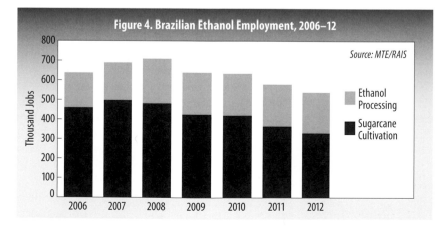

Figure 4. Brazilian Ethanol Employment, 2006–12

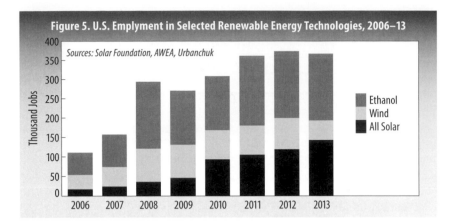

Figure 5. U.S. Emplyment in Selected Renewable Energy Technologies, 2006–13

rising feedstock prices, reduced yields due to drought, and lower demand.[45] In 2013, however, conditions improved and employment stabilized at the previous year's level.[46]

In most countries, the number of renewable energy jobs is still comparatively limited, and often there is simply no reliable information at all. (Even for India, one of the countries included in Table 1, the information is dated.) Among the noteworthy exceptions is Bangladesh, the developing world's leader in installations of so-called solar home systems (single-panel PV assemblies used mostly in rural areas). The number of such systems has climbed from 15,000 in 2003 to 2.8 million in 2013, with about 80,000 units newly installed each month.[47] This has increased the estimated number of jobs in the country from 60,000 in 2011 to more than 100,000 in 2013.[48] Japan, a leading PV manufacturer and installer, is estimated to have had a total of 60,000 jobs in 2012, according to the Japan Photovoltaic Energy Association.[49] This number likely increased significantly in 2013, when domestic shipments of solar modules more than doubled.[50] In Malaysia, which has become an important PV manufacturer, employment rose from 7,300 in 2012 to 9,200 jobs in 2013.[51]

Better information is necessary for a range of countries to generate a more complete and accurate renewable energy employment picture. Attention is also needed on the question of whether development of renewable energy leads to job loss elsewhere. For now, renewables are mostly supplementing conventional sources of energy, but they will eventually replace polluting fuels such as coal. For biofuels, food-versus-fuel trade-offs can translate into loss of livelihoods in agriculture.

All in all, available information suggests that renewable energy has grown to become a significant source of jobs. Rising labor productivity notwithstanding, the job numbers are likely to grow in coming decades as the world's energy system shifts toward low-carbon sources.

Chronic Hunger Falling,
But One in Nine People Still Affected

Gaelle Gourmelon

Although the proportion of people experiencing chronic hunger is decreasing globally, one in nine individuals still does not get enough to eat.[1] The United Nations Food and Agriculture Organization estimates that 805 million people were living with undernourishment (chronic hunger) in 2012–14, down more than 100 million over the last decade and 209 million lower than in 1990–92.[2] (See Figure 1.) The vast majority of undernourished people live in developing countries, where an estimated 791 million people—or one in eight—were chronically hungry in 2012–14.[3]

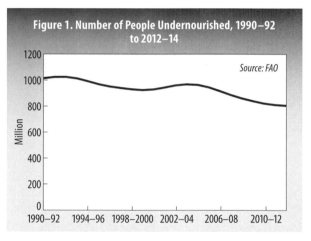

Figure 1. Number of People Undernourished, 1990–92 to 2012–14

Source: FAO

Undernourishment is defined as an inability to take in enough calories over at least one year to meet dietary energy requirements.[4] It can lead to *undernutrition*, a broader term that describes a condition caused by a deficient or imbalanced diet or by poor absorption and biological use of nutrients within the body.[5] Undernutrition can in turn lead to impaired physical functions and has high social and economic impacts, with the combined cost of undernutrition and micronutrient deficiencies equivalent to $1.4–$2.1 trillion per year, or 2–3 percent of gross world product.[6]

Women and children are particularly vulnerable to nutritional deficiencies due to biological and social inequities.[7] Women's low educational levels, unequal social status, and limited decision-making power can influence both their own nutritional status and that of their children. An undernourished mother is more likely to give birth to a low birth weight baby, causing an intergenerational cycle of poverty and undernutrition. Undernourished children are at higher risk of death from infectious diseases (like diarrhea and pneumonia) and can experience devastating physical, social, and economic consequences into adulthood.[8] Globally, undernutrition contributes to more than one-third of child deaths.[9]

The hunger target of Millennium Development Goal 1c (MDG-1c)—to halve the proportion of the population in developing countries who are hungry from the 1990 base year to the 2015 target year—is within reach.[10] Since 1990–92, the prevalence of chronic hunger fell from 18.7 percent to 11.3 percent in 2012–14, less than 2 percent above the MDG-1c target.[11] With 805 million people undernourished in 2012–14, however, the world is not on track to reach the more ambitious

Gaelle Gourmelon is the communications and marketing manager at Worldwatch Institute.

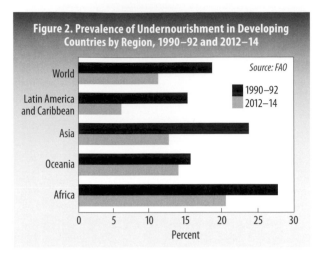

Figure 2. Prevalence of Undernourishment in Developing Countries by Region, 1990–92 and 2012–14

Source: FAO

■ 1990–92
▨ 2012–14

1996 World Food Summit target, which aimed to reduce the actual number of hungry people to 412 million by 2015 (from the 1996 baseline of 824 million).[12]

While there is global progress in reducing hunger, stark disparities exist between and within regions. (See Figure 2.). Many countries have made little or no progress.

Latin America and the Caribbean has shown the greatest reduction in undernourishment and has already achieved the MDG-1c target.[13] Since the early 1990s, the prevalence of chronic hunger in this region fell by almost two-thirds.[14] Haiti, with nearly half of its people undernourished, faces the largest burden in the region.[15]

Oceania had one of the lowest levels of chronic hunger at the beginning of the 1990s. However, the prevalence has fallen only 1.7 percentage points over the last two decades, leaving a higher share of the population undernourished in 2012–14 than in any other region except Africa.[16]

Northern Africa has consistently had a prevalence of undernourishment of less than 5 percent.[17] But the sub-Saharan region has by far the highest prevalence of any region. While the prevalence there has declined from 33.3 percent in 1990–92, one in four people is still chronically hungry.[18] Nearly half of the people in Zambia were undernourished in 2012–14.[19]

Asia, as a whole, is close to reaching the MDG-1c, with the prevalence of undernourishment decreasing from 23.7 percent in 1990–92 to 12.7 percent in 2012–14.[20] Yet because of Asia's large population, two out of three undernourished people in the world (526 million people) live in this region.[21] In West Asia, the prevalence of chronic hunger actually increased from 6.3 percent in 1990–92 to 8.7 percent in 2012–14 due to political and economic instability.[22]

Beyond looking at the prevalence of undernourishment, several measures of food security can point to underlying causes of undernutrition more broadly. The World Food Summit measures four dimensions of food security: availability, access, stability, and utilization.[23] Availability encompasses the quantity, quality, and diversity of food. Access means physical and economic access. Stability covers food security risks, like dependence on irrigation or food imports, and the incidence of shocks, such as fluctuations in domestic food supply or political instability. Utilization addresses the physiological ability to absorb and use food in the body; sanitation and health care are major contributors to this dimension.

Climate change is presenting an unprecedented challenge to all these dimensions due to disruptions in supply chains, increases in market prices, decreased assets and livelihood opportunities, lower purchasing power, and threats to human health.[24] The market sensitivity to climate change was highlighted recently by several periods of rapid increases in food prices following climate extremes, like heat waves, droughts, floods, cyclones, and wildfires, in key producing regions.[25] Food insecurity and the breakdown of food systems due to climate change particularly affect poorer populations.[26]

Because poverty is the main determinant of hunger, access to food is determined by incomes, food prices, and the ability to get social support.[27] Food prices have been fluctuating greatly, although generally rising since the late 1990s.[28] World food prices have begun to come down since the all-time peak in August 2012, but they remain high (see Figure 3), and poor households still spend a significant portion of their incomes on food.[29]

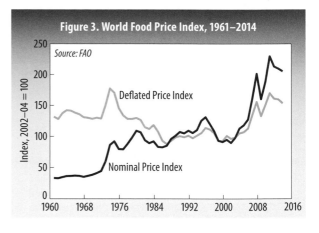

Due to income growth and poverty reduction in many countries, some progress on food access has been made over the last two decades.[30] Globally, food availability has also improved, with per capita food supply increasing from 2,200 kilocalories (kcal) per day in the early 1960s to more than 2,800 kcal by 2009.[31] Large challenges remain in food utilization and in stability due to food price fluctuations and natural and human-made crises. However, progress on addressing the various driving factors of food insecurity is uneven across developing regions.

In Latin America, economic growth, political stability, and agricultural and economic incentives have helped the region reach its hunger-reduction target. Utilization of food has improved thanks in part to better access to improved water sources and sanitation facilities.[32] In countries with social protection, such as safety nets, access to food has improved. Governments like Brazil have coordinated an array of policies with strong engagement from civil society, cutting undernourishment from 10.7 percent to less than 5 percent between 2000–02 and 2004–06.[33] The Caribbean, however, faces low stability of food security due to the heavy reliance on international food markets, low domestic food access, and limited natural resources.

In sub Saharan Africa, in contrast, high poverty rates, deteriorating rural infrastructure, and slow income growth mean food availability and distributional access remain low, leaving the region to struggle with the greatest food security challenge. Inadequate safe drinking water and sanitation facilities have limited people's ability to absorb and use the food that is available. As populations are displaced due to violence in the Central African Republic and Nigeria, increases in demand in certain areas of Chad and Niger have affected cereal prices there.[34] And in 2014, certain areas of West Africa experienced restricted trade flows and market disruptions due to the Ebola Virus Disease outbreak.[35] The effects of Ebola on food prices are not yet clear in Guinea, Liberia, and Sierra Leone, the most affected countries.[36]

Access to food has improved significantly in East and Southeast Asia. China, Indonesia, Thailand, and Vietnam have experienced rapid economic growth over the last 30 years, leading to greater investment in agriculture, more food availability, and better access to food.[37] In South Asia, however, social protection programs have not been enough to extend the benefits of regional economic growth to the poor.[38] In West Asia, political and economic instability, mainly due to conflict in Iraq (where the proportion chronically hungry rose dramatically from 7.9 percent in 1990–92 to 23.5 percent in 2012–14), have contributed to an increase in the prevalence of hunger.[39]

Food aid programs peaked in 2000–01.[40] The 1999 Food Aid Convention (FAC), a multilateral donor cooperation treaty that aimed to contribute to world food security, saw a drastic drop in annual food aid shipments from 10.5 million wheat ton equivalents in 2000–01 to 5.7 million wheat ton equivalents in 2011–12.[41] The United States provided the majority of international food aid, supplying 56 percent of food aid shipments from 1995–96 to 2011–12.[42]

The Food Assistance Convention, which replaced the expired FAC in 2013, includes not only commodities (like food and seeds) but also cash-based assistance.[43] Thanks to growing recognition that local and regionally purchased food aid is often faster and cheaper, the new treaty stresses that assistance should not require recipients to purchase food from the donor country.[44] The new treaty does, however, make the levels of assistance and the impact of international food price volatility more unpredictable. Unlike the 1999 treaty, the Food Assistance Convention does not establish minimum annual food commitments from donors. Instead, members now announce their commitment levels each year. The new treaty allows donors to express their commitments in terms of quantity of food or in value in the currency of their choice, making assistance vulnerable to international food price volatility.[45] The impacts of the Food Assistance Convention remain to be seen.

The fundamental human right to food, which is codified by the Universal Declaration of Human Rights, must be protected through social, economic, and political policies on food and health.[46] Through investments, sound policy making, strong legal frameworks, stakeholder involvement, and evidence-based decision making, the food security and nutrition environment can be improved to eradicate hunger worldwide.

Notes

Global Coal Consumption Keeps Rising, But Growth Is Slowing (pages 2–5)

1. BP, *Statistical Review of World Energy 2014* (London: 2014).

2. Ibid.

3. Ibid.

4. Ibid.; Matt Lucky and Reese Rogers, "Global Coal and Natural Gas Consumption Continue to Grow," in Worldwatch Institute, *Vital Signs, Vol. 20* (Washington, DC: Island Press, 2013), pp. 6–10; Janet L. Sawin, "Fossil Fuel Use Up," in Worldwatch Institute, *Vital Signs 2003* (New York: W. W. Norton & Company, 2003), pp. 34–35. BP data begin in 1965; data for years before are from Sawin.

5. BP, op. cit. note 1.

6. Ibid.

7. Ibid.

8. International Energy Agency, *World Energy Outlook 2014 Fact Sheet* (Paris: 2014).

9. Energy Information Administration (EIA), *International Energy Statistics—Countries* (Washington, DC: 2014).

10. Ibid.

11. Ibid. Conversion of short tons (mass unit) to mtoe (energy unit) based upon EIA data on 2013 U.S. coal consumption in short tons and BP data in tons of oil equivalent. The resulting conversion factor (0.49) is subject to uncertainties, as different sorts of coal can vary in the energy they contain per unit mass. General trends are not affected by this factor uncertainty.

12. BP, op. cit. note 1.

13. Ibid.

14. Ibid.

15. EIA, *What Is the Average Heat Content of U.S. Coal? Historical Average Annual Heat Content: Appendix* (Washington, DC: 2014); 1 Megajoule (MJ) = 0.278 kWh.

16. BP, op. cit. note 1.

17. Ibid.

18. Ibid.

19. Ibid.

20. EIA, *International Energy Statistics, Coal Imports by Region* (Washington, DC: 2014).

21. BP, op. cit. note 1.

22. Ibid.

23. Euan Mearns, *What Happens When Chinese Coal Production Stops Growing 10% Yearly?*, www.businessinsider.com, 19 November 2010.

24. Ibid.

25. EIA, *Countries – China – Analysis – Overview* (Washington, DC: 2014).

26. Du Juan, "China Coal Imports to Continue Affecting Global Prices: Platts," *China Daily*, 22 May 2013.

27. EIA, op. cit. note 25.

28. Giles Parkinson, "China Ramps Up 2015 Solar Target to 40 GW," *Reneweconomy.com*, 12 December 2012.

29. "China Tops 2011 Index Rankings for Renewable Energy," *RenewableEnergyWorld.com*, 23 August 2011; "Don't Let China Dominate," www.usnews.com, 18 September 2014, citing IEA data.

30. EIA, op. cit. note 20.

31. Center for Climate and Energy Solutions, *Energy and Climate Goals of China's 12th Five-Year Plan* (Arlington, VA: 2011)

32. EIA, *AEO2014 Early Release Overview* (Washington, DC: 2013); Jeffrey Logan, "U.S. Power Sector Undergoes Dramatic Shift in Generation Mix," National Renewable Energy Laboratory, 2013, at financere.nrel.gov.

33. BP, op. cit. note 1.

34. EIA, *Quarterly Coal Report* (Washington, DC: 2014)

35. Ibid.

36. BP, op. cit. note 1.

37. Ibid.

38. European Environmental Agency, *Trends in Energy*

Intensity, Gross Domestic Product and Gross Inland Energy Consumption (Copenhagen: 2014).

39. Ibid.

40. BP, op. cit. note 1.

41. "European Renewable Energy Share Increases to 14.4 % in 2012" (press releae), *EurObserv'ER*, December 2013.

42. European Commission, *EU Energy in Figures – Statistical Pocketbook 2013* (Brussels: 2013).

43. Ibid.

44. BP, op. cit. note 1; European Commission, op. cit. note 42.

45. "Hard Coal Imports into the EU-28 by Country of Origin," *Eurostat*, 2013; European Commission, op. cit. note 42.

46. German Federal Statistical Office, data and press releases; *Verein der Kohlenimporteure, Jahresbericht (annual report) 2013* and webpage (Hamburg: 2013).

47. Peter Frumhoff, "2° C or Not 2° C: Insights from the Latest IPCC Climate Report," Union of Concerned Scientists, 27 September 2013.

Wind Power Growth Still Surging Where Strongly Supported (pages 6–9)

1. Global Wind Energy Council (GWEC), *Global Wind Report Annual Market Update 2013* (Brussels: 2014).

2. Ibid.

3. Ibid.

4. Ibid.

5. Ibid.

6. Ibid.

7. REN21, *Renewables 2014 Global Status Report* (Paris: 2014).

8. Ibid.

9. Ibid.

10. Ibid.

11. Ibid.

12. Ibid.

13. Ibid.

14. Ibid.

15. Ibid.

16. Ibid.

17. GWEC, op. cit. note 1.

18. Ibid.

19. GWEC, "Wind Power in India Getting Back on Track," undated, at www.gwec.net/wind-power-india-getting-back-track.

20. Tildy Bayar, "Relief for India's Wind Industry as GBI Returns," *RenewableEnergyWorld.com*, 20 March 2013.

21. GWEC, op. cit. note 1.

22. Ibid.

23. Ibid.

24. Ibid.

25. Ibid.

26. Ibid.

27. Ibid.

28. "Wind-Energy Production Credit Breezes into Tax 'Extenders' Bill," *MarketWatch Wall Street Journal* (blog), 3 April 2014.

29. Ibid.

30. GWEC, op. cit. note 1.

31. Ibid.

32. Ibid.

33. Ibid.

34. Ibid.

35. Ibid.

36. Multilateral Investment Fund and Bloomberg New Energy Finance, "Climatescope," Washington, DC, 2013.

37. GWEC, op. cit. note 1.

38. REN21, op. cit. note 7.

39. Ibid.

40. U.S. Energy Information Administration, "Levelized Cost and Levelized Avoided Cost of New Generation Resources in the Annual Energy Outlook 2014," Washington, DC, 17 April 2014.

41. Silvio Marcacci, "Grid Parity, Low LCOE Driving 34% Global Renewables Capacity by 2030," *Clean Technica*, 18 October 2013.

42. Shaun Campbell, "Japan Plays the Long Game with Floating Technology," *Wind Power Monthly*, January 2014.

43. REN21, op. cit. note 7.

44. GWEC, op. cit. note 1.

45. Ibid.

46. Ibid.

47. Ibid.

48. Ibid.

Solar Power Installations Jump to a New Annual Total (pages 10–13)

1. While other important solar technologies exist—in particular, solar water heating—this Vital Sign highlights the latest trends in electricity generating markets.

2. REN21, *Renewables 2014 Global Status Report* (Paris: 2014).

3. Ibid.

4. Ibid.

5. Ibid.

6. Ibid. Figure calculated from REN21 data: PV installations at 39GW and CSP at 0.9GW; net capacity additions at 213.6 GW.

7. European Photovoltaic Industry Association (EPIA), *Global Market Outlook for Photovoltaics 2014–2018* (Brussels: 2014).

8. Frankfurt School–UNEP Centre/Bloomberg New Energy Finance, *Global Trends in Renewable Energy Investment 2014* (Frankfurt: 2014); REN21, op. cit. note 2. Data for the single years of 2004–10 are calculated as a two-year change from REN21 data.

9. Mike Munsell,"Solar Module Spot Price Hit Historic Lows in Q2 2014," *Greentech Solar*, 11 July 2014.

10. International Energy Agency (IEA), *PVPS Report: Snapshot of Global PV 1992–2013* (Paris: 2014).

11. REN21, op. cit. note 2.

12. Ibid.

13. Ibid.

14. EPIA, op. cit. note 7.

15. REN21, op. cit. note 2; BP, *Statistical Review of World Energy 2014* (London: 2014). Consumption data calculated from BP Energy Statistics.

16. BP, op. cit. note 15. Calculated from data.

17. REN21, op. cit. note 2.

18. James Arye, "Only 188 MW of Distributed Solar PV Installed in 1st Quarter of 2014," *Clean Technica*, 23 May 2014.

19. "PV Installations Grow Rapid in Japan," *PV Magazine*, 24 April 2014; "Japan Adds 4.58 GW of PV in Eight Months, FIT Cuts of 10% Mooted," *PV Magazine*, 4 March 2014.

20. "Japan Adds 4.58 GW," op. cit. note 19; Jonathan Gif-

ford, "Land of the Highly Prized Sun," *PV Magazine*, December 2013; "New World Solar Power Record for 1 Quarter, 2014 IHS Target Goes Up," *Clean Technica*, 10 April 2014.

21. REN21, op. cit. note 2.

22. EPIA, op. cit. note 7. Calculated from installations from 2012 to 2013: 17,726 MW to 10,975 MW.

23. Ibid.

24. Ibid.

25. REN21, op. cit. note 2. Other sources show that the United Kingdom installed less than 1 GW of PV.

26. IEA, op. cit. note 10; EPIA, op. cit. note 7.

27. IEA, op. cit. note 10; EPIA, op. cit. note 7.

28. BP, op. cit. note 15.

29. REN21, op. cit. note 2.

30. Solar Energy Industries Association (SEIA), *Solar Market Insight Report 2013 Year in Review* (Washington, DC: 2014).

31. BP, op. cit. note 15.

32. EPIA, op. cit. note 7.

33. BP, op. cit. note 15.

34. REN21, op. cit. note 2; EPIA, op. cit. note 7.

35. REN21, op. cit. note 2; EPIA, op. cit. note 7.

36. REN21, op. cit. note 2.

37. Ibid. SEIA reports U.S. installed 410 MW of CPS in 2013.

38. EurObserv'ER, *Solar Thermal and Concentrated Solar Barometer* (Brussels: 2014).

39. REN21, op. cit. note 2.

40. Ibid.

41. BP, op. cit. note 15.

42. Ibid.

43. Ibid.

44. Ibid. Calculated from data.

45. EurObserv'ER, *Photovoltaic Barometer 2014* (Brussels: 2014).

46. RED Eléctrica de España, *The Spanish Electricity System: Preliminary Report 2013* (Madrid: 2013). Does not include Spanish extra-peninsular territories; yearly data end on 17 December 2013.

47. Ibid.; U.S. Energy Information Administration, *Monthly Energy Review June 2014* (Washington, DC: 2014); Bruno Burger, *Electricity Production from Solar and Wind in Germany*

in 2013 (Freiburg, Germany: Fraunhofer Institute for Solar Energy Systems ISE, 2014); Terna, "Early Data on 2013 Electricity Demand: 317 Billion KWh of Demand, –3.4% Compared to 2012," press release (Rome: 9 January 2014).

48. Shyam Mehta, "Global 2013 PV Module Production Hits 39.8GW; Yingli Is the Shipment Leader," *GreenTech Media*, 23 April 2014.

49. Ibid.

50. Ibid.

51. Ibid.

52. Ibid.; IHS, "Leading Solar Module Suppliers Extend Dominance in 2013; Chinese Still on Top," press release (Munich: 30 April 2014).

53. Mehta, op. cit. note 48.

54. Ibid. This refers to conventional thin-film modules, with some exceptions.

55. Ibid.

56. REN21, op. cit. note 2.

57. Ibid.

58. EurObserv'ER, op. cit. note 45.

Wind, Solar Generation Capacity Catching Up with Nuclear Power (pages 14–17)

1. World Nuclear Association, "The Nuclear Renaissance," updated January 2014, at www.world-nuclear.org/info.

2. U.S. Energy Information Administration, *International Energy Outlook 2013* (Washington, DC: 2013), Figure 83.

3. Mycle Schneider et al., *The World Nuclear Industry Status Report 2014* (London: Mycle Schneider Consulting, 2014), p. 13.

4. U.S. Department of Energy, *2012 Renewable Energy Data Book* (Washington, DC: 2013), p. 48.

5. Ibid.

6. Ibid., p. 41.

7. BP, *Statistical Review of World Energy 2014* (London: 2014).

8. Ibid.

9. Ibid.

10. Schneider et al., op. cit. note 3, p. 6.

11. Ibid., p. 12.

12. Ibid.

13. International Atomic Energy Agency (IAEA), *Power Reactor Information System* (online database), "Nuclear Power Capacity Trend," at www.iaea.org/PRIS/WorldStatistics/WorldTrendNuclearPowerCapacity.aspx; Alexander Ochs and Michelle Ray, "Nuclear Power Recovers Slightly, But Global Future Uncertain," *Vital Signs Online*, 8 October 2013; BP, op. cit. note 7; Christopher Flavin, "Wind Energy Growth Continues," in Worldwatch Institute, *Vital Signs 2001* (New York: W. W. Norton & Company, 2001), p. 47.

14. Global Wind Energy Council (GWEC), *Global Wind Report Annual Market Update 2013* (Brussels: 2014).

15. REN21, *Renewables 2014 Global Status Report* (Paris: 2014).

16. IAEA, op. cit. note 13; Ochs and Ray, op. cit. note 13; BP, op. cit. note 7; Flavin, op. cit. note 13.

17. Author's calculation based on IAEA, op. cit. note 13, on Ochs and Ray, op. cit. note 13, on BP, op. cit. note 7, and on Flavin, op. cit. note 13.

18. BP, op. cit. note 7.

19. Calculated from ibid.

20. REN21, op. cit. note 15, p. 47.

21. Open Energy Information, "Transparent Cost Database," at en.openei.org/apps/TCDB.

22. Ibid.

23. Zachary Shahan, "Wind Turbine Net Capacity Factor—50% the New Normal?" *Clean Technica*, 27 July 2012.

24. International Energy Agency (IEA), *World Energy Investment Outlook* (Paris: 2014), Annex A.

25. Ibid.

26. Ibid.

27. Ibid.

28. International Renewable Energy Agency (IRENA), *Renewable Energy and Jobs* (Abu Dhabi: 2013).

29. Schneider et al., op. cit. note 3, p. 74; IRENA, op. cit. note 28.

30. GWEC, op. cit. note 14.

31. Ibid.

32. Based on data in IEA, "Data Services," online database, at wds.iea.org/WDS/Common/Login/login.aspx. Data include R&D for all energy purposes, not just the power sector.

33. Ibid.

34. Ibid.

35. Ibid.

36. Ibid.

37. REN21, op. cit. note 15, p. 115.

38. Ibid.

39. Schneider et al., op. cit. note 3, p. 13.

40. REN21, op. cit. note 15, p. 56.

41. Schneider et al., op. cit. note 3, p. 14.

42. Ibid, p. 73.

43. Ibid.

Smart Grid Investment Grows with Widespread Smart Meter Installations (pages 18–21)

1. Bloomberg New Energy Finance (BNEF), "China Outspends the US for First Time in $15bn Smart Grid Market," press release, 18 February 2014.

2. Ibid.

3. Ibid.

4. Ibid.

5. Charlene Fowler, "Smart Grid Market to Surpass $400 Billion Worldwide by 2020," *Greentech Media,* 13 August 2013.

6. Ibid.; Reese Rogers, "Smart Grid and Energy Storage Installations Rising," *Vital Signs Online* (Washington, DC: Worldwatch Institute, 27 February 2013).

7. Jeff St. John, "Report: China Outspent US on Smart Grid in 2013," *Greentech Media,* 20 February 2014.

8. Ibid.; BNEF, op. cit. note 1.

9. BNEF, op. cit. note 1.

10. U.S. Department of Energy (DOE), *Smart Grid Investment Grant Program: Progress Report II* (Washington, DC: October 2013), p. iv.

11. Ibid.

12. Ibid.

13. Innovation, Electricity, Efficiency (IEE), *Utility-scale Smart Meter Deployments: A Foundation for Expanded Grid Benefits* (Washington, DC: Edison Foundation, 2013), p. 1.

14. Ibid.

15. DOE, op. cit. note 10.

16. Eanna Kelly, "EU's Joint Research Centre (JRC) Counts 459 Smart Grid Initiatives in Its Yearly Stock-taking Report," *ScienceBusiness.net,* 27 March 2014.

17. Ibid.

18. Rogers, op. cit. note 6; Jesse Berst, "Smart Meters: What We're Learning in Europe (I Found #12 Most Intriguing)," *SmartGridNews.com,* 25 February 2014.

19. Berst, op.cit. note 18.

20. Ibid.

21. Stefan Nicola, "Germany Rejects EU Smart-Meter Recommendations on Cost Concerns," *Bloomberg News,* 1 August 2013.

22. Ibid.

23. Katherine Tweed, "Iberdrola Picks Winner(s) for 1 Million Smart Meters," *Greentech Media,* 19 March 2012.

24. Ibid.

25. Jeff St. John, "Europe's Smart Meter Race Hitting Its Stride," *Greentech Media,* 18 July 2013.

26. Ibid.

27. Ibid.

28. Ibid.

29. Government of the UK, "Helping Households to Cut Their Energy Bills," *Policy: Detail: Smart Meters,* London, updated 1 May 2014.

30. Ibid.

31. Department of Energy and Climate Change (DECC), *Smart Metering Implementation Programme: Second Annual Report on the Roll-out of Smart Meters* (London: 2013), p. 6.

32. DECC, *Smart Meters, Great Britain, Quarterly Report to End December 2013,* Statistical Release: Experimental National Statistics (London: 2014), p. 1.

33. Ibid.

34. Government of Japan, *Japan's First Biennial Report under the UNFCCC* (Tokyo: 2013), p. 51.

35. Rogers, op. cit. note 6.

36. "Japan's Power Utilities Accelerating Plans for Smart Meters," *Nikkei Asian Review,* 15 March 2014.

37. Ministry of Power, *Smart Grid Vision and Roadmap for India* (New Delhi: Government of India, 2013).

38. Ibid.

39. "Brazil Could Be an Unexpected Leader in Smart Meter Rollouts," *Gentrack,* 24 February 2014.

40. "South America Grid Market Forecast," *The Smart Grid Observer,* 11 June 2013.

41. Jesse Berst, "Brazil Ramps Up Smart Grid Efforts," *SmartGridNews.com,* 9 August 2013.

42. Norton Rose Fulbright, "The Inova Energia Program–Promoting Energy Efficiency and Renewable Energy in Brazil," at nortonrosefulbright.com, May 2013.

Global Energy and Carbon Intensity Continue to Decline (pages 22–25)

1. World Bank, GDP Indicators (in 2005 U.S. dollars), at data.worldbank.org/indicator/NY.GDP.MKTP.KD; BP, *Statistic Review of World Energy 2014* (London: 2014).

2. World Bank, op. cit. note 1; BP, op. cit. note 1.

3. World Bank, op. cit. note 1; BP, op. cit. note 1.

4. World Bank, op. cit. note 1; BP, op. cit. note 1.

5. World Bank, op. cit. note 1; BP, op. cit. note 1.

6. World Bank, op. cit. note 1; BP, op. cit. note 1.

7. World Bank, op. cit. note 1; BP, op. cit. note 1.

8. World Bank, op. cit. note 1; BP, op. cit. note 1.

9. World Bank, op. cit. note 1; BP, op. cit. note 1.

10. World Bank, GDP Indicators (in 2011 international dollars), at data.worldbank.org/indicator/NY.GDP.MKTP.PP .KD; BP, op. cit. note 1.

11. World Bank, op. cit. note 10; BP, op. cit. note 1.

12. World Bank, op. cit. note 10; BP, op. cit. note 1.

13. World Bank, op. cit. note 10; BP, op. cit. note 1.

14. World Bank, op. cit. note 10; BP, op. cit. note 1.

15. World Bank, op. cit. note 10; BP, op. cit. note 1.

16. World Bank, op. cit. note 10; BP, op. cit. note 1.

17. World Bank, op. cit. note 10; BP, op. cit. note 1.

18. World Bank, op. cit. note 10; BP, op. cit. note 1.

19. World Bank, op. cit. note 10; BP, op. cit. note 1.

20. World Bank, op. cit. note 10; BP, op. cit. note 1.

21. World Bank, op. cit. note 10; BP, op. cit. note 1.

22. World Bank, op. cit. note 10; Global Carbon Atlas, at www.globalcarbonatlas.org/?q=en/emissions.

23. World Bank, op. cit. note 10; *Global Carbon Atlas*, op. cit. note 22.

24. World Bank, op. cit. note 10; *Global Carbon Atlas*, op. cit. note 22.

25. World Bank, op. cit. note 10; *Global Carbon Atlas*, op. cit. note 22.

26. World Bank, op. cit. note 10; *Global Carbon Atlas*, op. cit. note 22.

27. World Bank, op. cit. note 10; *Global Carbon Atlas*, op. cit. note 22.

28. World Bank, op. cit. note 10; *Global Carbon Atlas*, op. cit. note 22.

29. "China Overtakes US as World's Biggest CO_2 Emitter," (London) *Guardian*, 19 June 2007.

30. World Bank, op. cit. note 10; *Global Carbon Atlas*, op. cit. note 22.

31. National Development and Reform Commission, "Climate Policies and Actions in China 2014 Annual Report" (in Mandarin Chinese), November 2014.

32. World Bank, op. cit. note 10; *Global Carbon Atlas*, op. cit. note 22.

33. The White House, "U.S.-China Joint Announcement on Climate Change," press release (Beijing: Office of the Press Secretary, 12 November 2014).

Effects and Sustainability of the U.S. Shale Gas Boom (pages 26–30)

1. "Shale Gas and Tight Oil Are Commercially Produced in Just Four Countries," *Today in Energy*, 13 February 2015. Note: Argentina, in spite of shale gas reserves, currently only produces tight oil.

2. Ibid.

3. Ibid.

4. Ibid.

5. "Shale Gas Provides Largest Share of U.S. Natural Gas Production in 2013," *Today in Energy*, 25 November 2014; U.S. Energy Information Administration (EIA), "Shale Gas Production," 4 December 2014, at www.eia.gov/dnav/ng /ng_prod_shalegas_s1_a.htm; EIA, International Energy Statistics, "Dry Natural Gas Production (2014)," at www .eia.gov/cfapps/ipdbproject/iedindex3.cfm?tid=3&pid=26 &aid=1&cid=regions&syid=1980&eyid=2013&unit=BCF.

6. "Shale Gas Provides Largest Share," op. cit. note 5; EIA, "Shale Gas Production," op. cit. note 5; EIA, "Dry Natural Gas Production," op. cit. note 5.

7. EIA, "Russia," 12 March 2014, at www.eia.gov/coun tries/cab.cfm?fips=rs.

8. EIA, *Technically Recoverable Shale Oil and Shale Gas Resources: An Assessment of 137 Shale Formations in 41 Countries Outside the United States* (Washington, DC: 2013).

9. Ibid.

10. "Shale Gas and Tight Oil," op. cit. note 1.

11. Ibid.

12. Ralf Dickel et al., *Reducing European Dependence on Russian Gas* (Oxford: Oxford Institute for Energy Studies, 2014); BP, *Statistical Review of World Energy 2014* (London: 2014).

13. Gail Tverberg, "Russia and the Ukraine—The Worri-

some Connection to World Oil and Gas Problems," *Ourfiniteworld.com*, 7 May 2014.

14. Emily Gosden and Tom Shiel, "Fracking: UK Shale Exploration Sites Mapped," (London) *Daily Telegraph*, 28 January 2015.

15. Ibid.

16. Ibid.

17. "Neue Schätzung stellt Fracking-Vorhaben in Frage," *Sueddeutsche Zeitung*, 9 January 2014.

18. Annett Meiritz and Anna Reimann, "Umstrittener Kabinettsbeschluss, Fracking Kommt—Hendricks Muss Damit Leben," *Spiegel Online*, 1 April 2015.

19. Eva Konzett and Aureliusz M. Pedziwol, "Chevron Gibt Osteuropa Endgültig Auf," *Wirtschaftsblatt*, 23 February 2015.

20. Ibid.

21. Ibid.

22. Ibid.

23. Anthony Fensom, "China: The New Shale-Gas Superpower?" *Nationalinterest.org*, 9 October 2014.

24. Ibid.

25. Ibid.

26. J. Robinson, "The Latin American Quandary: Lots of Shale Gas, Not a Lot of Production," *The Barrel*, 4 July 2014.

27. BP, op. cit. note 12.

28. Michael Ratner et al., *U.S. Natural Gas Exports: New Opportunities, Uncertain Outcomes* (Washington, DC: Congressional Research Service, 2015).

29. EIA, "U.S. Price of Natural Gas Delivered to Residential Customers," 31 March 2015, at www.eia.gov/dnav/ng/hist/n3010us3a.htm.

30. Ibid.

31. "European Union Natural Gas Import Price Chart," *Ycharts.com*, at ycharts.com/indicators/europe_natural_gas_price; "Multiple Factors Push Western Europe to Use Less Natural Gas and More Coal," *Today in Energy*, 27 September 2013. Note: Prices for consumers in EU member states vary from stated import prices.

32. "European Union Natural Gas," op. cit. note 31.

33. EIA, "AEO2014 Early Release Overview," 2014, p. 14, at www.eia.gov/forecasts/aeo/er/pdf/0383er%282014%29.pdf.

34. Ibid.

35. Ibid.

36. European Commission, *EU Energy in Figures—Statistical Pocketbook 2013* (Brussels: 2013); "Hard Coal Imports into the EU-28 by Country of Origin," *Eurostat*, 2013.

37. BP, op. cit. note 12.

38. See, for example, Dave Messersmith, "Understanding the Role of Liquefied Natural Gas; Part 1" (blog), Penn State University Extension, State College, PA, 19 February 2012; Xun Yao Chen, "Why Liquefaction Costs Will Affect Liquefied Natural Gas Trade Growth," *Market Realist*, 27 May 2014. Note: 1 cubic foot = 1,020 BTU; data apply to U.S. liquefied natural gas.

39. BP, op. cit. note 12.

40. Maria Gallucci, "Feds Approve Fourth LNG Export Terminal Amid Growing Pressure to Cash in on US Energy Boom," *International Business Times*, 30 September 2014.

41. Ibid.

42. Ibid.

43. "Liquefied Natural Gas—Bubbling Up," *The Economist*, 31 May 2014.

44. BP, op. cit. note 12.

45. Ibid.

46. Jane Perlez, "China and Russia Reach 30-Year Gas Deal," *New York Times*, 22 May 2014.

47. Donald Gilliland, "US Chamber of Commerce Launches Pro-gas Campaign with Inaccurate Jobs Numbers," *Harrisburg Patriot-News*, 19 July 2012; Frank Mauro et al., *Exaggerating Shale Drilling's Employment Impacts: How & Why* (Harrisburg, PA: Multi-State Shale Research Collaborative, 2013).

48. U.S. Department of Labor, Bureau of Labor Statistics, "Economy at a Glance—Pennsylvania," viewed 2 March 2015; Pennsylvania Department of Labor and Industry, "Marcellus Shale Fast Facts," Harrisburg, PA, January 2014 and November 2014.

49. Marcellus Shale Education & Training Center, *Marcellus Shale Workforce Needs Assessment* (Williamsport, PA: 2010), p. 22.

50. Marie Cusick, "Gas Industry Survey Shows Job Growth Slowing," *State Impact*, 29 July 2014.

51. Stephen Lacey, "MA Has Double the Jobs in Clean Energy That PA Has in Natural Gas," *GreenTechMedia.com*, 19 September 2013.

52. Mike Lee, "Drilling Boom Costs Pa. Thousands per Well in Road Damage," *E&E Publishing*, 27 March 2014; Elizabeth Ridlington and John Rumpler, *Fracking by the Numbers: Key Impacts of Dirty Drilling at the State and National Level* (Washington, DC: Environment America, October 2013);

Tony Dutzik, Elizabeth Ridlington, and John Rumpler, *The Costs of Fracking: The Price Tag of Dirty Drilling's Environmental Damage* (Washington, DC: Environment America Research and Policy Center, 2012).

53. EIA, "Short-Term Energy Outlook, Table 7c: U.S. Regional Electricity Prices (Cents per Kilowatthour)," at www.eia.gov/forecasts/steo/tables/?tableNumber=21#startcode=1996.

54. David Hughes, *Fracking Fracas: The Trouble with Optimistic Shale Gas Projections by the U.S. Department of Energy* (Santa Rosa, CA: Post Carbon Institute, 2014).

55. Ibid.; Mason Inman, "Natural Gas: The Fracking Fallacy," *Nature*, 3 December 2014; Hannah Petersen, "Shale Gas Projections Are in Decline—And We Shouldn't Be Surprised," *TheConversation.com*, 6 December 2014.

Greenhouse Gas Increases Are Leading to a Faster Rate of Global Warming (pages 32–35)

1. Global Carbon Project, "Global Carbon Budget 2014," Power Point Presentation, 21 September 2014.

2. Ibid.

3. Fiona Harvey, "Record CO_2 Emissions 'Committing the World to Dangerous Climate Change'," (London) *Guardian*, 21 September 2014.

4. "CO_2 Emissions Set to Reach New 40 Billion Tonne Record High in 2014," press release (Norwich, U.K.: Tyndall Center for Climate Research, 22 September 2014).

5. International Energy Agency, "About Climate Change," at www.iea.org/topics/climatechange.

6. Steven Munson, "All Over the Planet, Countries are Completely Missing Their Emissions Targets" (blog), *Washington Post*, 23 September 2014.

7. U.N. Environment Programme, "UN Says Global Carbon Neutrality Should be Reached by Second Half of Century, Demonstrates Pathways to Stay Under 2°C Limit," press release (Nairobi: 19 November 2014).

8. Ibid.

9. Global Carbon Project, op. cit. note 1.

10. Stefan Nicola, "China Surpasses EU in Per-Capita Pollution for First Time," *Bloomberg News*, 22 September 2014.

11. Global Carbon Project, op. cit. note 1.

12. Harvey, op. cit. note 3.

13. Joe Romm, "Methane Leaks Wipe Out Any Benefit of Fracking, Satellite Observations Confirm," *Thinkprogress.org*, 22 October 2014.

14. Ibid.

15. U.S. Environmental Protection Agency (EPA), "Global Greenhouse Gas Emissions Data," at www.epa.gov/climatechange/ghgemissions/global.html, based on IPCC data.

16. Grass reduces belching in ruminant livestock compared with grains and therefore reduces the production of methane.

17. REN21, *Renewables Global Status Report 2014* (Paris: 2014), p. 15.

18. Ray Massey and Hannah McClure, "Agriculture and Greenhouse Gas Emissions," Commercial Agriculture Program, University of Missouri Extension Service, 2014.

19. Ibid.

20. EPA, "National Greenhouse Gas Emissions Data," at www.epa.gov/climatechange/ghgemissions/usinventoryreport.html.

21. "UN Climate Summit Pledges to Halt Loss of Natural Forests by 2030," (London) *Guardian*, 23 September 2014.

22. Ibid.

23. REN21, op. cit. note 17, p. 28.

24. Ibid., p. 68.

25. Ibid., p. 48.

26. Katie Sullivan, "Private Sector Banking Could Transform the UN's Green Climate Fund," *Responding to Climate Change*, 13 October 2014.

Global Coastal Populations at Risk as Sea Level Continues to Rise (pages 36–38)

1. Commonwealth Scientific and Industrial Research Organisation, *Reconstructed GMSL for 1880 to 2009* (Australia: 2011).

2. Ibid.

3. Ibid.

4. J. A. Church and N. J. White, "Sea-Level Rise from the Late 19th to the Early 21st Century," *Surveys in Geophysics*, September 2011, pp. 585–602.

5. Gordon McGranahan, Deborah Balk, and Bridget Anderson, "The Rising Tide: Assessing the Risks of Climate Change and Human Settlements in Low Elevation Coastal Zones," *Environment and Urbanization*, April 2007, pp. 17–37.

6. Intergovernmental Panel on Climate Change (IPCC), *Climate Change 2013: The Physical Science Basis* (Cambridge, U.K.: Cambridge University Press, 2013), p. 17.

7. National Oceanic and Atmospheric Administration (NOAA), Earth System Research Laboratory, "Trends in Atmospheric Carbon Dioxide," at www.esrl.noaa.gov/gmd/ccgg/trends.

8. IPCC, op. cit. note 6, p. 37.

9. Ibid., p. 162.

10. Ibid., p. 167.

11. Ibid., p. 8.

12. National Aeronautics and Space Administration, *Global Climate Change: Evidence*, at climate.nasa.gov/evidence.

13. IPCC, op. cit. note 6, p. 9.

14. Ibid.

15. Ibid., p. 320.

16. Ibid., p. 9.

17. Ibid., p. 49.

18. Ibid., p. 24.

19. Ibid., p. 49.

20. Ibid., p. 25.

21. Ibid.

22. NOAA, *National Coastal Population Report: Population Trends from 1970 to 2020* (Washington DC: 2013), p. 3.

23. McGranahan, Balk, and Anderson, op. cit. note 5.

24. Ibid.

25. Avery Fellow, "Natural Disasters Expected to Pose Greater Risk to Global Supply Chain, Study Says," *Bloomberg BNA*, August 2012.

26. McGranahan, Balk, and Anderson, op. cit. note 5.

27. Daniel Pruzin, "2013 Set to be One of the Warmest Years on Record, UN Meteorological Agency Says," *Bloomberg BNA*, November 2013.

28. Ibid.

29. Ibid.

30. NYC Department of City Planning, Current Population Estimates, at www.nyc.gov/html/dcp/html/census/popcur.shtml.

31. PlaNYC, *A Stronger, More Resilient New York* (New York: 2013).

32. IPCC, *Climate Change 2014: Impacts, Adaptation, and Vulnerability* (Geneva: 2014), p. 25.

Auto Production Sets New Record, Fleet Surpasses 1 Billion Mark (pages 40–44)

1. Colin Couchman, IHS Automotive, London, e-mail to author, 27 May 2014.

2. Ibid.

3. Ibid.

4. Ibid.

5. Ibid.

6. Ibid.

7. Ibid.

8. Ibid.

9. Ibid. People per vehicle calculated from data in Population Reference Bureau, *2013 World Population Data Sheet* (Washington, DC: 2014).

10. Couchman, op. cit. note 1.

11. Stacy C. Davis, Susan W. Diegel, and Robert G. Boundy, *Transportation Energy Data Book: Edition 32* (Oak Ridge, TN: Center for Transportation Analysis, Oak Ridge National Laboratory, 2013), Table 3-6, p. 3-10.

12. Ibid., p. 3-7.

13. Ibid., p. 3-8.

14. Michael Sivak, "Has Motorization in the U.S. Peaked? Part 5: Update through 2012," University of Michigan Transportation Research Institute, April 2014, Table 1.

15. Ibid.

16. Ibid., Tables 3–5.

17. Michael Sivak, "Has Motorization in the U.S. Peaked? Part 4: Households without a Light-Duty Vehicle," University of Michigan Transportation Research Institute, January 2014, Table 1.

18. Ibid., Table 2.

19. Davis, Diegel, and Boundy, op. cit. note 11, Table 8-16, p. 8-20.

20. Ibid.

21. Ibid., Table 3-2, p. 3-3.

22. European Environment Agency (EEA), *Monitoring CO$_2$ Emissions from New Passenger Cars in the EU: Summary of Data for 2013* (Copenhagen: 2014).

23. Davis, Diegel, and Boundy, op. cit. note 11, Table 3-2, p. 3-3. Estimates for national and global car fleets vary, substantially so in some cases. China's National Bureau of Statistics, for instance, suggests that the country has 73.26 million registered private cars, far more than the 43.2 million estimated by IHS in 2011. However, this article relies on a single source for consistency of estimates across countries.

24. Ibid.

25. Ibid.

26. Ibid.

27. Global Fuel Economy Initiative, "Fuel Economy and the UN's Post 2015 Sustainable Development Goals," 2014, at www.globalfueleconomy.org/Documents/fuel-economy-and-un-post-2015-sdg.pdf.

28. Ibid.

29. International Council on Clean Transportation (ICCT), "Global Passenger Vehicle Standards Update. February 2014 Datasheet," Washington, DC.

30. EEA, op. cit. note 22.

31. Ibid.

32. Ibid.

33. Ibid.

34. ICCT, *European Vehicle Market Statistics 2013* (Berlin: 2013), pp. 70–71.

35. Ibid.

36. Michael Sivak, "Making Driving Less Energy Intensive than Flying," University of Michigan Transportation Research Institute, January 2014, Table 1.

37. Ibid.

38. ICCT, op. cit. note 34, p. 88.

39. Ibid.

40. EEA, op. cit. note 22.

41. Ibid.

42. Ibid.

43. Green Car Congress, "Navigant Forecasts Hybrids to Account for Almost 4% of Global Light-duty Vehicles Sales by 2020, Plug-ins 3%," 11 June 2013.

44. Ibid.

45. Electric Drive Transportation Association, "Electric Drive Sales Dashboard," at electricdrive.org/index.php?ht=d/sp/i/20952/pid/20952, viewed 10 May 2014.

46. Davis, Diegel, and Boundy, op. cit. note 11, Table 6-4, p. 6-7.

47. Toyota, "Worldwide Sales of Toyota Hybrids Top 6 Million Units," 15 January 2014, at corporatenews.pressroom.toyota.com/releases/worldwide+toyota+hybrid+sales+top+6+million.htm?view_id=35924.

48. Ibid.

49. Ibid.

50. IHS, "Global Production of Electric Vehicles to Surge by 67 Percent This Year," 4 February 2014, at press.ihs.com/press-release/automotive/global-production-electric-vehicles-surge-67-percent-year.

51. Green Car Congress, "IHS Automotive Forecasts Global Production of Plug-in Vehicles to Rise by 67% This Year," 4 February 2014 at www.greencarcongress.com/2014/02/20140204-ihspev.html.

52. Zentrum für Sonnenenergie- und Wasserstoff-Forschung Baden-Württemberg, "Weltweit über 400.000 Elektroautos unterwegs," press release (Ulm, Germany: 31 March 2014).

53. Ibid.

Passenger and Freight Rail Trends Mixed, High-Speed Rail Growing (pages 45–49)

1. International Union of Railways (UIC), "Synopsis 2013," Paris, June 2014. Note that the UIC estimates for any year are typically composed of a range of years, since the statistics for some countries are not as up-to-date as for others. The passenger rail data in this Vital Sign do not include urban systems such as subways, light rail, and streetcars.

2. Ibid.; UIC, earlier annual Synopsis reports, at www.uic.org/spip.php?rubrique1449 and www.uic.org/spip.php?article1350.

3. UIC, op. cit. note 1; 2000–13 data from UIC, op. cit. note 2; 1980–99 data from World Bank Railways Database, cited in Molly O. Sheehan, "Passenger Rail at a Crossroads," in Worldwatch Institute, *Vital Signs 2002* (New York: W. W. Norton & Company, 2002), p. 79.

4. UIC, op. cit. note 1.

5. Calculated from ibid. and from UIC, op. cit. note 2.

6. UIC, op. cit. note 1. A time series from 1980 comparable to that for passenger rail is not available due to gaps in World Bank data, which render the resulting global numbers incompatible with UIC data.

7. UIC, "Synopsis 2000," Excel database file, at www.uic.org/IMG/xls/synth2000.xls.

8. UIC, op. cit. note 2.

9. Ibid.

10. UIC, op. cit. note 1.

11. UIC, "Synopsis 2001," at www.uic.org/IMG/xls/synth2001.xls.

12. Stacy C. Davis, Susan W. Diegel, and Robert G. Boundy, *Transportation Energy Data Book*, Edition 33 (Oak Ridge, TN: Center for Transportation Analysis, Oak Ridge National Laboratory, July 2014), Table 9-8.

13. Ibid.

14. UIC, op. cit. note 1.

15. UIC, "Synopsis 1991," at www.uic.org/IMG/xls/Syn these_1991.xls; UIC, op. cit. note 11.

16. Stephen Joiner, "Is Bigger Better? 'Monster' Trains vs Freight Trains," *Popular Mechanics*, 11 February 2010; Wikipedia, "Longest Trains," at en.wikipedia.org/wiki/Longest_trains.

17. UIC, op. cit. note 1.

18. Ibid.

19. Ibid.

20. Ibid.

21. UIC, op. cit. note 15.

22. Calculated from UIC, op. cit. note 1, and from UIC, op. cit. note 7.

23. Calculated from UIC, op. cit. note 1, and from UIC, op. cit. note 7. Europe's share includes Russia.

24. UIC, op. cit. note 1.

25. Calculated from UIC, op. cit. note 1, and from UIC, op. cit. note 7.

26. Calculated from UIC, op. cit. note 1, and from UIC, op. cit. note 7.

27. UIC, op. cit. note 1.

28. Calculated from UIC, op. cit. note 1, and from UIC, op. cit. note 7.

29. Calculated from UIC, op. cit. note 1, and from UIC, op. cit. note 7.

30. Richard Freeman and Hal Cooper, "Why Electrified Rail Is Superior," *21st Century Science & Technology*, Vol. 18, No. 2 (2005).

31. Calculated from UIC, op. cit. note 1.

32. UIC, op. cit. note 1.

33. Ibid.

34. Ibid.

35. Calculated from ibid., and from UIC, op. cit. note 2.

36. UIC, op. cit note 1.

37. Ibid.

38. Davis, Diegel, and Boundy, op. cit. note 12, Table 2-12.

39. Ibid., Table 2-15.

40. Oil Change International, *Runaway Train: The Reckless Expansion of Crude-by-Rail in North America* (Washington, DC: May 2014).

41. Ibid.

42. Ibid.

43. Jad Mouawad, "U.S. Issues Safety Alert for Oil Trains," *New York Times*, 7 May 2014.

Aquaculture Continues to Gain on Wild Fish Capture (pages 52–55)

1. U.N. Food and Agriculture Organization (FAO), "World Fish Trade to Set New Records," press release (Rome: 21 February 2014).

2. FAO Fisheries and Aquaculture Department "Global Capture Production Statistics 2012," at ftp://ftp.fao.org/FI/STAT/Overviews/CaptureStatistics2012.pdf.

3. FAO, *The State of the World Fisheries and Aquaculture 2012* (Rome: 2012).

4. Ibid.; FAO, op. cit. note 2; FAO Fisheries and Aquaculture Department, "FAO Global Aquaculture Production Volume and Value Statistics Database Updated to 2012," March 2014, at ftp://ftp.fao.org/FI/STAT/Overviews/AquacultureStatistics2012.pdf.

5. FAO, op. cit. note 1.

6. World Bank, *Fish to 2030: Prospects for Fisheries and Aquaculture* (Washington, DC: December 2013).

7. Ibid.

8. FAO, "Fact Sheet: International Fish Trade and World Fisheries," February 2014, at ftp://ext-ftp.fao.org/FI/Data/cofi_ft/COFI_FT_Factsheet.pdf.

9. FAO, op. cit. note 4.

10. Ibid.

11. FAO, op. cit. note 2.

12. Ibid.

13. Ibid.

14. Ibid.

15. Ibid.

16. Ibid.

17. FAO, op. cit. note 4.

18. Ibid.

19. Ibid.

20. Ibid.

21. FAO, op. cit. note 8.

22. Ibid.

23. Ibid.

24. Ibid.

25. Ibid.

26. Ibid.

27. Ibid.

28. Ibid.

29. Ibid.

30. Ibid.

31. World Bank, op. cit. note 6.

32. FAO, op. cit. note 1.

33. Ibid.

34. FAO, op. cit. note 3, p. 41.

35. Ibid., p. 42.

36. Ibid., p. 41.

37. Ibid.

38. Ibid.

39. Ibid., p. 46.

40. Calculated from data in Population Reference Bureau, *2012 World Population Data Sheet* (Washington, DC: 2012).

41. FAO, op. cit. note 1.

42. Ibid.

43. FAO, op. cit. note 3, p. 46.

44. Ibid.

45. Ibid., p. 47.

46. Ibid.

47. World Ocean Review, "The Global Hunt for Fish," at worldoceanreview.com/en/wor-2/fisheries/state-of-fisheries-worldwide.

48. Ibid.

49. World Ocean Review, "Plenty More Fish in the Sea?" at worldoceanreview.com/en/wor-2/fisheries.

50. International Labour Organization, *Working Towards Sustainable Development: Opportunities for Decent Work and Social Inclusion in a Green Economy* (Geneva: 2012).

51. Ibid.

52. Craig Leisher, "Milestone Looms for Farm-Raised Fish," Green Blog, *New York Times*, 24 January 2013.

53. Ibid.; Marc Gunther, "Shrimp Farms' Tainted Legacy Is Target of Certification Drive," *Environment 360*, 6 August 2012; Aquaculture Stewardship Council, at www.asc-aqua.org; Global Aquaculture Alliance, "Best Aquaculture Practices," at www.gaalliance.org.

54. FAO, op. cit. note 3, p. 191.

55. World Bank, op. cit. note 6.

Peak Meat Production Strains Land and Water Resources (pages 56–59)

1. U.N. Food and Agriculture Organization (FAO), *Food Outlook*, May 2014, p. 7.

2. Ibid.

3. Ibid., pp. 45–49.

4. Growth rate in 2010 from Laura Reynolds and Danielle Nierenberg, "Disease and Drought Curb Meat Production and Consumption," in Worldwatch Institute, *Vital Signs, Vol. 20* (Washington, DC: Island Press, 2013), p. 49.

5. FAO, *FAOSTAT Statistical Database*, at faostat.fao.org; FAO, op. cit. note 1.

6. Calculated from Doug Boucher et al., *Grade A Choice? Solutions for Deforestation-Free Meat* (Cambridge, MA: Union of Concerned Scientists, 2012), p. 2, from U.N. Department of Economic and Social Affairs, *World Population to 2300* (New York: 2004), and from Population Reference Bureau, *2013 World Population Data Sheet* (Washington, DC: 2013).

7. FAO, op. cit. note 1, p. 111.

8. Ibid.

9. Ibid.

10. Ibid.

11. Ibid.

12. Ibid.

13. Heinrich Böll Stiftung (HBS) and Friends of the Earth Europe (FoEE), *Meat Atlas: Facts and Figures About the Animals We Eat* (Berlin and Brussels, 2014), pp. 12–13.

14. Calculated from FAO, op. cit. note 5, and from FAO, op. cit. note 1, p. 7.

15. FAO, op. cit. note 1, p. 111.

16. Ibid.

17. Ibid.

18. Ibid.

19. Ibid., p. 7.

20. Ibid.

21. Ibid.

22. FAO, op. cit. note 5.

23. Calculated from ibid.

24. Ibid.

25. FAO, "FAO Food Price Index," at www.fao.org/world foodsituation/foodpricesindex/en.

26. Ibid.

27. Ibid.

28. FAO, op. cit. note 1, p. 7.

29. Ibid.

30. Ibid.

31. Ibid., pp. 112–15.

32. Calculated from ibid., p. 114.

33. Calculated from ibid., p. 113.

34. See, for instance, HBS and FoEE, op. cit. note 13.

35. Ibid., p. 26.

36. Ibid.

37. Boucher et al., op. cit. note 6, p. 2.

38. Ibid., p. 7.

39. Ibid., p. 1.

40. HBS and FoEE, op. cit. note 13, p. 31.

41. Ibid., p. 28.

42. Table 1 from Water Footprint, "Animal Products," at www.waterfootprint.org/?page=files/Animal-products.

43. HBS and FoEE, op. cit. note 13, p. 29.

Coffee Production Near Record Levels, Prices Remain Volatile (pages 60–63)

1. U.S. Department of Agriculture, Foreign Agricultural Service (USDA-FAS), "Table 01 Coffee World Production, Supply and Distribution," database at apps.fas.usda.gov /psdonline/psdHome.aspx, created 20 June 2014. The data cover production of green (unroasted) coffee. USDA-FAS measures production in units of 60-kilogram bags; conversion to tons by author.

2. USDA-FAS, "Coffee: World Markets and Trade. 2014/15 Forecast Overview," June 2014, at apps.fas.usda.gov/psdon line/circulars/coffee.pdf.

3. USDA-FAS, op. cit. note 1.

4. U.N. Food and Agriculture Organization (FAO), FAO-STAT Statistical Database, at faostat3.fao.org/faostat-gateway /go/to/download/Q/QC/E, viewed 19 July 2014.

5. USDA-FAS, op. cit. note 1; FAO, op. cit. note 4.

6. FAO, op. cit. note 4.

7. Ibid.

8. Ibid.

9. Ibid.

10. USDA-FAS, "Table 03A Coffee Production," at apps.fas .usda.gov/psdonline, created 20 June 2014.

11. Ibid.

12. Data for 2013/14 from ibid; top 10 in 2000 from Brian Halweil, "Coffee Production Hits New High," in World-watch Institute, Vital Signs 2001 (New York: W. W. Norton & Company, 2001), p. 36.

13. Halweil, op. cit. note 12.

14. Calculated from USDA-FAS, op. cit. note 1.

15. Ibid.

16. USDA-FAS, "Table 04 Coffee Consumption," at apps .fas.usda.gov/psdonline, created 20 June 2014. USDA-FAS reports data in units of thousands of 60-kilogram bags; conversion to tons by author.

17. Ibid.

18. Ibid.

19. U.N. Conference on Trade and Development, UNCTAD-STAT online database, commodity trends, at unctadstat.unc tad.org/wds/TableViewer/tableView.aspx?ReportId=28768, viewed 26 July 2014.

20. Ibid.

21. Noemie Eliana Maurice and Junior Davis, Unravelling the Underlying Causes of Price Volatility in World Coffee and Cocoa Commodity Markets, Discussion Paper 1, UNCTAD Special Unit on Commodities (Geneva: 2011), p. 8.

22. Ibid., p. 4.

23. Ibid., p. 6.

24. Number of coffee farmers from Fairtrade International, "Problems Facing Coffee Producers," at www.fairtrade.net /coffee.html, viewed 26 July 2014.

25. Maurice and Davis, op. cit. note 21, p. 6.

26. U.N. Development Programme, Towards Human Resilience: Sustaining MDG Progress in an Age of Economic Uncertainty (New York: 2011), p. 73.

27. Jason Potts, Michael Opitz, and Chris Wunderlich, Closing the Gaps in GAPS: A Preliminary Appraisal of the Measures and Costs Associated with Adopting Commonly Recognized "Good Agricultural Practices" in Three Coffee Growing Regions (Winnipeg, Canada: International Institute for Sustainable Development, 2007), p. 4.

28. International Trade Centre, Trends in the Trade of Certified Coffees (Geneva: 2011).

29. Ibid.

30. Ibid.

31. Ibid.

32. Ibid.

33. Ibid.

34. International Labour Organization (ILO), *Working Towards Sustainable Development. Opportunities for Decent Work and Social Inclusion in a Green Economy* (Geneva: 2012).

35. Oromia Coffee Farmers Cooperative Union, "Welcome to Oromia Coffee Farmers Cooperative Union," at www.oro miacoffeeunion.org, viewed 26 July 2014.

36. Ibid.

37. Ibid.

38. ILO, *Sustainable Development, Decent Work and Green Jobs*, International Labour Conference, Report V (Geneva: 2013).

Volatile Cotton Sector Struggles to Balance Cost and Benefits (pages 64–67)

1. Share in 1940s from Fairtrade International, "Cotton," at www.fairtrade.net/cotton.html; 1960 from U.N. Conference on Trade and Development (UNCTAD), "Cotton Market," 28 February 2011, at www.unctad.info/en/Infocomm /Agricultural_Products/Cotton/Market.

2. U.N. Food and Agriculture Organization (FAO) and International Cotton Advisory Committee (ICAC), *World Apparel Fiber Consumption Survey* (Washington, DC: 2013).

3. Number of households from Fairtrade International, op. cit. note 1; number of countries from Better Cotton Initiative, "Key Facts," at bettercotton.org/wp-content/up loads/2014/11/BCI-Key-Facts-20141.pdf.

4. ICAC, *Cotton This Month*, 5 January 2015. The ICAC reporting period for cotton statistics stretches from August of one year to July of the next.

5. ICAC, *Cotton: World Statistics Bulletin of the International Cotton Advisory Committee* (Washington, DC: 2012).

6. Calculated from ICAC, op. cit. note 4, and from ICAC, op. cit. note 5.

7. ICAC, op. cit. note 4; ICAC, op. cit. note 5.

8. ICAC, op. cit. note 5, p. 11.

9. ICAC, op. cit. note 4; ICAC, op. cit. note 5; ICAC, *Cotton This Month*, 2 June 2014; ICAC, *Cotton This Month*, 3 January 2014.

10. ICAC, op. cit. note 4, p. 5.

11. Data for 2013/14 calculated from ICAC, op. cit. note 4, pp. 5–6; 1949/50 calculated from ICAC, op. cit. note 5, p. 3.

12. Data for 2014 calculated from ICAC, op. cit. note 4, pp. 5–6, and from U.S. Department of Agriculture, Foreign Agricultural Service (USDA-FAS), "Cotton: World Markets and Trade," January 2015.

13. Based on data in ICAC, op. cit. note 4, and on ICAC, op. cit. note 5.

14. Cotton Incorporated, "China: Center of the Cotton Market," undated, at www.cottoninc.com/corporate/Market -Data/SupplyChainInsights/China-Cotton-Market-01-11; James Johnson et al., "The World and United States Cotton Outlook," presentation at Agricultural Outlook Forum 2014, Arlington, VA, 21 February 2014.

15. ICAC, op. cit. note 4.

16. Ibid.; ICAC, op. cit. note 5.

17. ICAC, op. cit. note 4; ICAC, op. cit. note 5.

18. James Kiawu, "China's Cotton Policies to Lower Domestic Consumption and Imports," USDA Economic Research Service, 1 April 2013.

19. ICAC, op. cit. note 4; ICAC, op. cit. note 5.

20. Based on data in ICAC, op. cit. note 4, on ICAC, op. cit. note 5, and on USDA-FAS, op. cit. note 12.

21. ICAC, op. cit. note 4, pp. 4, 5.

22. Fairtrade Foundation, *The Great Cotton Stitch-Up* (London: November 2010), p. 11.

23. Rural population share from ibid.; export dependence from UNCTAD, op. cit. note 1.

24. UNCTAD, op. cit. note 1.

25. Ibid.

26. "Cotton Monthly Price," *Index Mundi*, at www.index mundi.com/commodities/?commodity=cotton&months =360.

27. ICAC, op. cit. note 5, p. 11.

28. Stocks from ICAC, op. cit. note 4; subsidies from Fairtrade International, op. cit. note 1.

29. Fairtrade Foundation, op. cit. note 22, p. 5.

30. International Centre for Trade and Sustainable Development (ICTSD), *Cotton: Trends in Global Production, Trade and Policy* (Geneva: 2013), pp. 4–5.

31. Pesticide Action Network North America, "Cotton," at www.panna.org/resources/cotton.

32. "The Risks of Cotton Farming," at www.organiccotton .org/oc/Cotton-general/Impact-of-cotton/Risk-of-cotton -farming.php.

33. Inyoung Hwan et al., "How Fair Trade Helped Indian Cotton Farmers When Government Failed," *Triple Pundit*, 2 May 2013.

34. U.S. Environmental Protection Agency, "Pesticides: Health and Safety. Human Health Issues," 17 October 2014, at www.epa.gov/pesticides/health/human.htm.

35. Hwan et al., op. cit. note 33.

36. Arjen Y. Hoekstra and Ashok K. Chapagain, "Water Footprints of Nations: Water Use by People as a Function of Their Consumption Pattern," *Water Resources Management*, vol. 21, no. 1 (2007), pp. 35–48.

37. "The Risks of Cotton Farming," op. cit. note 32; Rebecca Lindsey, "Shrinking Aral Sea," NASA Earth Observatory, 25 August 2000.

38. Ashok K. Chapagain et al., "The Water Footprint of Cotton Consumption: An Assessment of the Impact of Worldwide Consumption of Cotton Products on the Water Resources in the Cotton Producing Countries," *Ecological Economics*, vol. 60 (2006), pp. 186–203.

39. Ibid.

40. Ibid.

41. Ibid.

42. Ibid.

43. Ibid.

44. Ibid.

45. Mesfin M. Mekonnen and Arjen Y. Hoekstra, "Water Footprint Benchmarks for Crop Production: A First Global Assessment," *Ecological Indicators*, No. 46 (November 2014), p. 220.

46. Field to Market, *Environmental and Socioeconomic Indicators for Measuring Outcomes of On-Farm Agricultural Production in the United States: Second Report, (Version 2)*, December 2012, available at www.fieldtomarket.org.

47. Cotton Australia, "Water," at cottonaustralia.com.au/cotton-library/fact-sheets/cotton-fact-file-water.

48. Mekonnen and Hoekstra, op. cit. note 45, p. 220.

49. Ibid.

50. Chapagain et al., op. cit. note 38.

51. Fairtrade International, op. cit. note 1.

52. "Organic Cotton," at www.organiccotton.org/oc/Organic-cotton/Organic-cotton.php.

53. Fairtrade International, op. cit. note 1.

54. Ibid.

55. "The Advantages of Fairtrade Cotton," undated, at www .organiccotton.org/oc/Fairtrade-cotton/Benefits-of-fairtrade-cotton.php.

56. Better Cotton Initiative, op. cit. note 3.

57. WWF, "Sustainable Agriculture: Cotton," at www.worldwildlife.org/industries/cotton.

58. Better Cotton Initiative, op. cit. note 3.

Genetically Modified Crop Industry Continues to Expand (pages 68–71)

1. C. James, *Global Status of Commercialized Biotech/GM Crops: 2014*, ISAAA Brief No. 49 (Ithaca, NY: International Service for the Acquisition of Agri-biotech Applications (ISAAA), 2014).

2. Ibid.

3. Ibid.

4. Ibid.

5. C. James, *Global Status of Transgenic Crops*, *Global Review of Commercialized Transgenic Crops*, *Global Status of Commercialized Transgenic Crops*, and *Global Status of Commercialized Biotech/GM Crops* (Ithaca, NY: ISAAA, 1997–2014).

6. C. James, *Global Review of Commercialized Transgenic Crops: 1998*, ISAAA Brief No. 8 (Ithaca, NY: ISAAA, 1998).

7. James, op. cit. note 1.

8. Ibid.

9. Ibid.

10. Ibid.

11. Ibid.

12. Y. Zheng, *Research, Deployment and Safety Management of Genetically Modified Poplars in China: Forests and Genetically Modified Trees* (Rome: U.N. Food and Agriculture Organization (FAO), 2010).

13. U.S. Department of Agriculture (USDA), *Information Systems for Biotechnology*, electronic database, at www.isb.vt.edu.

14. C. James, *Global Status of Commercialized Biotech/GM Crops: 2012*, ISAAA Brief No. 44 (Ithaca, NY: ISAAA, 2012).

15. C. James, *Global Status of Commercialized Biotech/GM Crops: 2013*, ISAAA Brief No. 46 (Ithaca, NY: ISAAA, 2013).

16. FAO, *FAOSTAT Statistical Database* (Rome: 2014).

17. James, op. cit. note 1.

18. Ibid.

19. Ibid.

20. Ibid.

21. Ibid.

22. Ibid.

23. Ibid.

24. USDA, op. cit. note 13.

25. J. Fernandez-Cornejo et al., *Genetically Engineered Crops in the United States* (Washington, DC: USDA, Economic Research Service, 2014); James, op. cit. note 1.

26. N. Makoni and J. Mohamed-Katerere, "Genetically Modified Crops," in U.N. Environment Programme, *Africa Environment Outlook 2. Our Environment, Our Wealth* (Nairobi: 2006).

27. N. Gilbert, "Case Studies: A Hard Look at GM Crops," *Nature*, 2 May 2013.

28. James, op. cit. note 1.

29. Ibid.

Food Trade and Self-Sufficiency (pages 72–76)

1. U.S. Department of Agriculture (USDA), *Production, Supply, and Distribution*, electronic database, at www.fas.usda.gov/psdonline, viewed 25 February 2015.

2. Ibid.

3. Ibid.

4. Worldwatch calculation based on data in USDA, note 1.

5. Ibid.

6. Ibid.

7. Marianela Fader et al., "Spatial Decoupling of Agricultural Production and Consumption: Quantifying Dependences of Countries on Food Imports due to Domestic Land and Water Constraints," *Environmental Research Letters*, 26 March 2013.

8. USDA, op. cit. note 1.

9. Stacey Rosen, Economic Research Service, USDA, email to author, 9 December 2014.

10. USDA, *Production, Supply, and Distribution*, electronic database, at apps.fas.usda.gov/psdonline, viewed 28 February 2015.

11. Fader et al., op. cit. note 7.

12. U.N. Food and Agriculture Organization (FAO), "Water Withdrawal by Sector, Around 2006," *Aquastat*, at www.fao.org/nr/aquastat, updated December 2012.

13. Arjen Y. Hoekstra and Mesfin M. Mekonnen, "The Water Footprint of Humanity," *Proceedings of the National Academy of Sciences*, 28 February 2012.

14. Ibid.

15. Ibid.

16. Ibid.

17. Ibid.; Arjen Y. Hoekstra, "Water Security of Nations: How International Trade Affects National Water Scarcity and Dependency," in J. Anthony Jones, Trahel Vardanian, and Christina Hakopian, eds., *Threats to Global Water Security* (NATO Science for Peace and Security Series C: Environmental Security) (Springer, 2009), pp. 27–36.

18. Stephen Foster and Tushaar Shah, *Groundwater Resources and Irrigated Agriculture: Making a Beneficial Relation More Sustainable* (Stockholm: Global Water Partnership, 2012); U.N. Environment Programme, "A Glass Half Empty: Regions at Risk Due to Groundwater Depletion," January 2012, at www.unep.org/pdf/UNEP-GEAS_JAN_2012.pdf; M. Falkenmark, "Growing Water Scarcity in Agriculture: Future Challenge to Global Water Security," *Philosophical Transactions of the Royal Society A*, September 2013; Amanda Mascarelli, "Demand for Water Outstrips Supply," *Nature*, 8 August 2012.

19. Katalyn A. Voss et al., "Groundwater Depletion in the Middle East from GRACE with Implications for Transboundary Water Management in the Tigris-Euphrates-Western Iran Region," *Water Resources Research*, February 2013, pp. 904–14.

20. Cynthia Barnett, "Groundwater Wake-up," *Ensia* (University of Minnesota), 19 August 2013.

21. Arjen Y. Hoekstra et al., "Global Monthly Water Scarcity: Blue Water Footprints versus Blue Water Availability" *PLOS ONE*, 29 February 2012.

22. Ibid.

23. FAO, *The State of the World's Land and Water Resources for Food and Agriculture* (Rome: 2011); Nikos Alexandratos and Jelle Bruinsma, *World Agriculture Towards 2030/2050* (Rome: FAO, 2012), p. 105.

24. FAO, op. cit. note 23.

25. Ibid.

26. USDA, *Summary Report: 2010 National Resources Inventory* (Washington, DC, and Ames, Iowa: Natural Resources Conservation Service and Center for Survey Statistics and Methodology, Iowa State University, 2013).

27. Land Matrix, electronic database, at landmatrix.org/en/get-the-idea/web-transnational-deals, viewed 26 February 2015; Japan area from U.S. Central Intelligence Agency, "Country Comparison: Area," in *The World Factbook*, at www.cia.gov/library/publications/the-world-factbook/rankorder/2147rank.html, viewed 13 August 2014.

28. Land Matrix, op. cit. note 27.

29. Ibid.

30. Ibid.

31. Ibid.

32. Ibid.

33. Lorenzo Cotulo, *Land Deals in Africa: What Is in the Contracts?* (London: International Institute for Environment and Development, 2011); Brian Bienkowski and Environmental Health News, "Corporations Grabbing Land and Water Overseas," *Scientific American*, 12 February 2013.

34. Land Matrix, op. cit. note 27.

35. Ibid.

36. Bienkowski and Environmental Health News, op. cit. note 33; Ward Anseeuw et al., *Land Rights and the Rush for Land: Findings of the Global Commercial Pressures on Land Research Project* (Rome: International Land Coalition, 2012).

37. Richard Howitt et al., *Economic Analysis of the 2014 Drought for California Agriculture* (Davis, CA: Center for Watershed Sciences, University of California, July 2014); 5 percent is a Worldwatch calculation based on total irrigated farmland in California from U.S. Geological Survey, at pubs.usgs.gov/circ/2004/circ1268/htdocs/table07.html.

38. Hoekstra and Mekonnen, op. cit. note 13.

Global Economy Remained a Mixed Bag in 2013 (pages 78–81)

1. International Monetary Fund (IMF), *World Economic Outlook Database 2014* (Washington, DC: 2014).

2. Ibid.

3. United Nations, *2013 World Economic Situation and Prospects* (New York: 2013), p. 4.

4. Global Economy Watch, *Business as Usual Is Changing: Our Predictions for 2013* (PricewaterhouseCoopers, 2013).

5. Paul Krugman and Maurice Obstfeld, *International Economics: Theory & Policy* (New York: Addison-Wesley, 2009).

6. Tim Callen, "PPP vs. the Market: Which Weight Matters," *Finance and Development,* March 2007; Krugman and Obstfeld, op. cit. note 5.

7. IMF, *World Economic Outlook 2013* (Washington, DC: 2013), p. xv.

8. Global Economy Watch, op. cit. note 4.

9. Ibid.

10. IMF, op. cit. note 1.

11. IMF, op cit. note 7.

12. D'Vera Cohn and Paul Taylor, *Baby Boomers Approach 65—Glumly* (Washington, DC: Pew Research Center, 2010).

13. Diana Farrell et al., *Talkin' 'Bout My Generation: The Economic Impact of Aging US Baby Boomers* (McKinsey Global Institute, 2008), p. 9.

14. The Economist, *Age Invaders* (London: 2014).

15. Ibid.

16. Ibid.

17. Sarah Marsh and Holger Hansen, "Insight: The Dark Side of Germany's Jobs Miracle," *Reuters*, 8 February 2012.

18. International Labour Organization (ILO), *Global Employment Trends 2014: Risk of a Jobless Recovery?* (Geneva: 2014), p. 11.

19. Richard B. Freeman, "Labor Market Imbalances: Shortages, Surpluses, or What?" in Jane Sneddon Little, ed., *Global Imbalances and the Evolving World Economy* (Boston: Federal Reserve Bank of Boston, 2008), pp. 159–82.

20. ILO, *Labour Market Policies and Institutions* (Geneva: 2014).

21. ILO, op. cit. note 18.

22. Ibid.

23. Ibid, p. 16.

24. Ibid.

25. Cohn and Taylor, op. cit. note 12.

26. ILO, op. cit. note 18, p. 16.

27. Ibid, p. 21.

28. Ibid, p. 11.

29. IMF, op. cit. note 7.

30. Kathy Bergen, "Jobs Coming Back Post-recession, But with Much Lower Pay, Study Says," *Chicago Tribune*, 11 August 2014.

31. U.N. Development Programme, *Humanity Divided: Confronting Inequality in Developing Countries* (New York: 2013), p. 94.

32. World Bank, "Gini Index," at data.worldbank.org/indicator/SI.POV.GINI.

33. Clean Techies, "August 19th Is Earth Overshoot Day," at cleantechies.com/2014/08/19/august-19th-is-earth-overshoot-day.

34. Ibid.

35. Herman E. Daly, *Beyond Growth: The Economics of Sustainable Development* (Boston: Beacon, 1997).

36. Paul Collier, *The Bottom Billion: Why the Poorest Countries*

Are Failing and What Can Be Done About It (Oxford: Oxford University Press, 2007), p. 14.

37. Ibid.

38. U.N. News Centre, "World Population Projected to Reach 9.6 Billion by 2050," at www.un.org/apps/news/story .asp?NewsID=45165#.VBxLMlduW4o.

39. Lew Daly and Stephen Posner, *Beyond GDP: New Measures for a New Economy* (New York: Demos, 2001), p. 8.

40. Ibid.

41. John Helliwell, Richard Layard, and Jeffrey Sachs, eds., *World Happiness Report* (New York: Sustainable Development Solutions Network et al., 2013), p. 3.

42. Ibid, p. 17.

43. Bruce Stokes, "Happiness Is Increasing in Many Countries—But Why?" Global Attitudes Project, Pew Research, 27 July 2007.

44. Jon Clifton, "People Worldwide Are Reporting a Lot of Positive Emotions," *Gallup*, 21 May 2014.

Commodity Prices Kept Slowing in 2013 But Still Strong Overall (pages 82–85)

1. Calculation by Worldwatch Institute based on *World Bank Annual Commodity Price Index 2014*, at go.worldbank .org/4ROCCIEQ50.

2. Morgan Stanley, *Commodities Super-Cycle: Is It Coming to An End?* (New York, 2013).

3. Bilge Erten and Jose A. Ocampo, *Super-Cycles of Commodity Prices Since the Mid-Nineteenth Century* (New York: United Nations Economic and Social Affairs, 2012), p.1.

4. Otaviano Canuto, *The Commodity Super Cycle: Is This Time Different?* (Washington, DC: World Bank Group, 2014), p. 1.

5. Ralph Atkins, "Falling Commodity Prices Flash Warning on Widening Global Divergences," *Financial Times*, 25 September 2014; Joe Richter, "Commodities 'Super Cycle' Is Seen Enduring by McKinsey," *Bloomberg*, 23 September 2013; Robin Harding and Michael Mackenzie, "Fed Renews Pledge on Low Rates," *Financial Times*, 17 September 2014.

6. Maria Kolesnikova, "Commodities Trading Jumped 23% Last Year Even as Banks Retreat," *Bloomberg*, 10 March 2014.

7. Ibid.

8. Ibid.

9. Francisco Costa, Jason Gerrad, and Joao Paulo Pessoa, *Winners and Losers from a Commodities-for-Manufactures*

Trade Boom (London: Center for Economic Performance, May 2014).

10. John Baffes, Cosic Camir, and Varun Kshirsaga, *Global Economic Prospects: Commodity Markets Outlook* (Washington, DC: World Bank, 2014), p. 2.

11. Ibid., p. 5.

12. Ibid., p. 1.

13. Richter, op. cit. note 5.

14. Martin Weil et al., *China's Energy Markets: Anhui, Chongqing, Henan, Inner Mongolia, and Guizhou Provinces* (Washington, DC: Environmental Protection Agency, Coalbed Methane Outreach Program, 2012), p. 13.

15. Ibid., p. 17.

16. Worldwatch Institute calculation, op. cit. note 1.

17. Baffes, Camir, and Kshirsaga, op. cit. note 10, p. 2.

18. Ibid.

19. Dorothy Kosich, "Reported Death of Commodities Super Cycle Premature," *Mineweb*, 20 September 2013.

20. Ibid.; Danske Bank, *Commodities 2014: Five Themes to Drive the Markets This Year* (Copenhagen: Danske Bank, 2014), p. 8.

21. China and Latin America, *Where Is China's Metal Industry Headed?* (Washington, DC: Inter-American Dialogue, 14 October 2014).

22. Tatyana Shumsky, "Gold Falls 28% in 2013, Ends 12-Year Bull Run," *Wall Street Journal*, 31 December 2013.

23. Ibid.

24. Ibid.

25. Richard Dobbs et al., *Resource Revolution: Tracking Global Commodity Markets* (McKinsey Global Institute, September 2013), p. 3.

26. "China's Food Security Dilemma" (blog), World Policy Institute, 4 June 2014.

27. Ibid.

28. Dobbs et al., op. cit. note 25, p. 3.

29. Ibid.

30. Ibid., p. 29.

Paper Production Levels Off (pages 86–90)

1. U.N. Food and Agriculture Organization (FAO), *FAOSTAT-Forestry Database*, at faostat3.fao.org/download/F/FO/E.

2. Ibid.

3. Calculated from ibid.

4. Ibid.

5. Ibid.

6. Ibid.

7. Ibid.

8. Ibid.

9. Ibid.

10. Ibid.

11. Ibid.

12. Ibid.; subsidies from Usha C. V. Haley, *No Paper Tiger: Subsidies to China's Paper Industry from 2002–09*, Briefing Paper No. 264 (Washington, DC: Economic Policy Institute, 2010).

13. FAO, op. cit. note 1. The estimate covers the years between 2002 and 2009.

14. United Nations Industrial Development Organization (UNIDO), "Strategy to Reduce Unintentional Production of POPs in China. Sino-Italian Project Funded for UNIDO POPs Program. Sector Report on Pulp & Paper Industry 2005," at www.unido.org/fileadmin/user_media /Services/Environmental_Management/Stockholm_Conven tion/POPs/CasestudyChina_PP_edited_by_Sbrilli_21_Dec _20071.pdf.

15. FAO, op. cit. note 1.

16. Ibid.

17. Ibid.

18. Ibid.

19. Deloitte, *2013 Global Forest, Paper, and Packaging Trend Watch—A Changing Landscape: South America's Influence on Global Markets* (2014).

20. FAO, op. cit. note 1.

21. Calculated from ibid.

22. Ibid.

23. Calculated from ibid.

24. Calculated from ibid.

25. Calculated from ibid.

26. Graeme Rodden, Mark Rushton, and Annie Zhu, "PPI Top 100: A Newcomer from Brazil is Welcomed to the List Along with Two from China," *PPI Magazine*, 1 September 2014. No production data are available for nine companies.

27. Ibid.

28. Calculated from ibid.

29. PricewaterhouseCoopers, *Global Forest, Paper & Packaging Industry Survey, 2014 Edition* (2014).

30. Jonathan Brandt, "Paper Demand Stacks Up," HSBC, 11 August 2014.

31. Ibid.; consumption shares in 2012 from Confederation of European Paper Industries, *European Pulp and Paper Industry Key Statistics 2013* (Brussels: 2014), p. 14.

32. FAO, *FAO Yearbook of Forest Products 2012* (Rome: 2014), pp. 186–88.

33. Product Stewardship Institute, "Garbage In, Garbage Out: Stopping the Unwanted Flow of Junk Mail," infographic at www.productstewardship.us/?Junk_Mail.

34. U.S. Environmental Protection Agency (EPA), "Paper Making and Recycling," Washington, DC, 14 November 2012.

35. EPA, "Recycling, Energy Conservation, and Community Beautification," at www.epa.gov/region3/beyond translation/2013BTF/SessionB-Beautification/Michelle Feldman.pdf.

36. Ibid.

37. FAO, "Global Production and Trade of Forest Products in 2013," 19 December 2013, at www.fao.org/forestry/statis tics/80938/en/. Percentage calculated from FAO, op. cit. note 1.

38. Percentage calculated from FAO, op. cit. note 1.

39. FAO, op. cit. note 37.

40. FAO, "Forest Product Consumption and Production," 24 December 2014, at www.fao.org/forestry/statistics/80 938@180723/en/.

41. Ibid.

42. Calculated from FAO, op. cit. note 1.

43. Calculated from ibid.

44. Calculated from ibid.

45. Calculated from ibid.

46. U.S. Energy Information Administration, Manufacturing Energy Consumption Survey, *2010 MECS Survey Data*, Table 1.2 at www.eia.gov/consumption/manufacturing /data/2010/#r1.

47. Natural Resources Defense Council, "Avoiding Chlorine in the Paper Bleaching Process," 20 September 2006, at www.nrdc.org/cities/living/chlorine.asp.

48. Environmental Paper Network (EPN), "Social Impacts of the Paper Industry," July 2007, at www.environmental paper.org; If You Care, "Why We Use Unbleached Totally Chlorine-Free (TCF) Paper," at www.ifyoucare.com/why

-we-use-unbleached-totally-chlorine-free-tcf-paper; Chinese companies from UNIDO, op. cit. note 14.

49. "Frequently Asked Questions on Kraft Pulp Mills," Ensis/CSIRO (Australia) joint research, 3 March 2005, at web.archive.org/web/20071202204438/http://www.gunn spulpmill.com.au/factsheets/BleachingByCSIRO.pdf. Kraft processes are used for most chemical pulping. Chemical pulping is the most commonly used process. In Europe, it accounts for two-thirds of total pulp production. See European Paper & Packaging Industries, "Types of Pulping Processes," undated, at www.paperonline.org/paper-making /paper-production/pulping/types-of-pulping-processes.

50. Haley, op. cit. note 12, pp. 8–9.

51. Deloitte, op. cit. note 19, p. 6.

52. EPN, *The State of the Paper Industry 2011. Steps Toward an Environmental Vision* (2011), at environmentalpaper.org /wp-content/uploads/2012/02/state-of-the-paper-industry -2011-full.pdf.

Global Plastic Production Rises, Recycling Lags (pages 91–94)

1. PlasticsEurope, *Plastics–The Facts 2014: An Analysis of European Plastics Production, Demand and Waste Data* (Brussels: 2014).

2. Ibid.

3. U.N. Environment Programme (UNEP), *Valuing Plastics: The Business Case for Measuring, Managing and Disclosing Plastic Use in the Consumer Goods Industry* (Nairobi: 2014).

4. Ibid.

5. Germany Trade & Invest, *Industry Overview: The Plastics Industry in Germany* (Berlin: 2014); PlasticsEurope, op. cit. note 1; American Chemistry Council, Economics and Statistics Department, *Plastic Resins in the United States* (Washington, DC: 2013).

6. American Chemistry Council, op. cit. note 5.

7. The European House–Ambrosetti, *The Excellence of the Plastics Supply Chain in Relaunching Manufacturing in Italy and Europe* (Milan: 2013); UNEP, op. cit. note 3.

8. American Chemistry Council, op. cit. note 5.

9. PlasticsEurope, op. cit. note 1; European House–Ambrosetti, op. cit. note 7.

10. American Chemistry Council, op. cit. note 5.

11. World Packaging Organization, *Market Statistics and Future Trends in Global Packaging* (Naperville, IL: 2009).

12. American Chemistry Council, Economics and Statistics Department, *Plastics and Polymer Composites in Light Vehicles* (Washington, DC: 2014).

13. Ibid.

14. First Research, *Plastic Resin & Synthetic Fiber Manufacturing Industry Profile* (Austin, TX: 2014).

15. European House–Ambrosetti, op. cit. note 7.

16. Germany Trade & Invest, op. cit. note 5.

17. Ibid.

18. PlasticsEurope, op. cit. note 1.

19. European House–Ambrosetti, op. cit. note 7.

20. Ibid.

21. PlasticsEurope, op. cit. note 1; American Chemistry Council, op. cit. note 5.

22. PlasticsEurope, op. cit. note 1.

23. Ibid.; American Chemistry Council, op. cit. note 5.

24. PlasticsEurope, op. cit. note 1; American Chemistry Council, op. cit. note 5.

25. American Chemistry Council, op. cit. note 5.

26. UNEP, op. cit. note 3.

27. Ibid.

28. Ibid.

29. PlasticsEurope, op. cit. note 1.

30. Ibid.

31. Ibid.

32. Ibid.

33. Ibid.

34. U.S. Environmental Protection Agency, *Plastics* (updated 2014), at www.epa.gov/osw/conserve/materials/plas tics.htm.

35. Ibid.

36. Costas Velis, *Global Recycling Markets: Plastic Waste; A Story for One Player—China* (Vienna: International Solid Waste Association, 2014).

37. UNEP, op. cit. note 3.

38. Velis, op. cit. note 36.

39. Ibid.

40. Ibid.

41. Ibid.

42. Ibid.

43. Ibid.

44. Ibid.

45. Ibid.

46. Ibid.

47. UNEP, op. cit. note 3.

48. Ibid.

49. Ibid.

50. Marcus Eriksen et al., "Plastic Pollution in the World's Oceans: More than 5 Trillion Plastic Pieces Weighing over 250,000 Tons Afloat at Sea," PLOS ONE, 10 December 2014.

51. Ibid.

52. UNEP, op. cit. note 3.

53. Eriksen et al., op. cit. note 50.

54. UNEP, op. cit. note 3.

55. Eriksen et al., op. cit. note 50.

56. Ibid.

57. UNEP, op. cit. note 3.

58. Ibid.

59. Clean Production Action, Plastics Scorecard v 1.0. Evaluating the Chemical Footprint of Plastics (Somerville, MA: 2014).

60. UNEP, op. cit. note 3.

61. Ibid.

62. Ibid.

63. European House–Ambrosetti, op. cit. note 7.

64. Ibid.

Will Population Growth End in This Century? (pages 96–99)

1. Patrick Gerland et al., "World Population Stabilization Unlikely This Century," Science, 10 October 2014, pp. 234–37; data are available at U.N. Population Division (UNPD) website, at esa.un.org/unpd/ppp; Wolfgang Lutz, William P. Butz, and Samir KC, eds., Executive Summary: World Population & Human Capital in the Twenty-First Century (Vienna and Laxenburg, Austria: Wittgenstein Centre for Demography and Global Human Capital and International Institute for Applied Systems Analysis (IIASA), 2014); data are available at Wittgenstein Centre website, at witt.null2.net/shiny/wittgensteincentredataexplorer.

2. Population Reference Bureau (PRB), "Human Population: Population Growth," at www.prb.org/Publications/Lesson-Plans/HumanPopulation/PopulationGrowth.aspx; UNPD, World Population Prospects: The 2012 Revision, File POP/1-1: "Total Population (Both Sexes Combined) by Major Area, Region and Country, Annually for 1950–2100 (Thousands)." (Growth rate calculations by the authors.)

3. UNPD, op. cit. note 2.

4. Gerland et al., op. cit. note 1; Lutz, Butz, and KC, op. cit. note 1.

5. John R. Wilmoth, "The U.N.'s Population Projections Are Likely to Be Right" (letter), Wall Street Journal, 27–28 September 2014; Wolfgang Lutz et al., "Population Growth: Peak Probability" (letter), Science, 31 October 2014, p. 562.

6. Gerland et al., op. cit. note 1.

7. Ibid.

8. Lutz, Butz, and KC, op. cit. note 1.

9. Ibid.

10. The Demographic and Health Surveys that shaped the U.N. assumptions about future fertility can be found at www.dhsprogram.com/publications/index.cfm.

11. UNPD, "Total Population (Both Sexes Combined) by Country or Area, 2010–2100 (Thousands)," at esa.un.org/unpd/ppp/Data-Output/UN_PPP2012_output-data.htm.

12. See, for example, "UNICEF Report: Africa's Population Could Hit 4 Billion by 2100," National Public Radio, 13 August 2014.

13. Lutz, Butz, and KC, op. cit. note 1.

14. Ibid.

15. UNPD website, op. cit. note 1; Wittgenstein website, op. cit. note 1.

16. UNPD website, op. cit. note 1; Wittgenstein website, op. cit. note 1.

17. UNPD website, op. cit. note 1; Wittgenstein website, op. cit. note 1.

18. UNPD website, op. cit. note 1; Wittgenstein website, op. cit. note 1.

19. UNPD website, op. cit. note 1; Wittgenstein website, op. cit. note 1.

20. U.S. Census Bureau, "World Population," at www.census.gov/population/international/data/worldpop/table_population.php; PRB, World Population Data Sheet 2014 (Washington, DC: 2014).

21. Lutz, Butz, and KC, op. cit. note 1; PRB, op. cit. note 20.

22. UNPD, op. cit. note 2.

23. Corey J. A. Bradshaw and Barry W. Brook, "Human Population Reduction is Not a Quick Fix for Environmental Problems," Proceedings of the National Academy of Sciences, 18 November 2014, pp. 16610–15.

24. Ibid.

25. Ibid.

26. Gilda Sedgh, Susheela Singh, and Rubina Hussain, "Intended and Unintended Pregnancies Worldwide in 2012 and Recent Trends," *Studies in Family Planning*, vol. 45, no. 3 (2014), pp. 301–14.

Jobs in Renewable Energy Expand in Turbulent Process (pages 100–04)

1. International Renewable Energy Agency (IRENA), *Renewable Energy and Jobs: Annual Review 2014* (Abu Dhabi: 2014).

2. IRENA, *Renewable Energy and Jobs* (Abu Dhabi: 2013).

3. United Nations Environment Programme (UNEP), *Green Jobs: Towards Decent Work in a Sustainable, Low-Carbon World* (Nairobi: 2008); International Labour Organization (ILO), *Working Towards Sustainable Development: Opportunities for Decent Work and Social Inclusion in a Green Economy* (Geneva: 2012).

4. Important methodology questions are discussed in IRENA, op. cit. note 2.

5. Ibid.

6. Ibid.

7. Ibid.

8. Ibid.

9. Charlie Zhu, "China's Solar Industry Rebounds, But Will Boom-Bust Cycle Repeat?" *Reuters*, 23 January 2014; Stephen Lacey, "China May Have Deployed More Solar in 2013 Alone than America Has Installed Altogether," *Greentechmedia*, 24 January 2014.

10. IRENA, op. cit. note 1.

11. Ibid.

12. UNEP, op. cit. note 3.

13. Ibid.

14. IRENA, op. cit. note 1.

15. EurObserv'ER, *Wind Energy Barometer* (Paris: 2014).

16. Ibid.

17. European Wind Energy Association, "Wind in Power: 2013 European Statistics," Brussels, February 2014.

18. IRENA, op. cit. note 1.

19. Werner Weiss and Franz Mauthner, *Solar Heat Worldwide: Markets and Contribution to the Energy Supply 2011* (Gleisdorf, Austria: International Energy Agency, Solar Heating & Cooling Programme, 2013).

20. China from Wang Zhongying, China National Renewable Energy Centre, e-mail to Arslan Khalid, IRENA, 12 April 2014; global figure from IRENA, op. cit. note 1.

21. IRENA, op. cit. note 1.

22. Ibid.

23. Calculated from Population Reference Bureau, *2013 World Population Data Sheet* (Washington, DC: 2013).

24. Zhongying, op. cit. note 20.

25. Ibid.

26. EurObserv'ER, *The State of Renewable Energies in Europe 2013* (Paris: 2014).

27. Ibid.

28. Ibid.; EurObserv'ER, op. cit. note 15.

29. EurObserv'ER, op. cit. note 26.

30. Marianne O'Sullivan et al., "Bruttobeschäftigung durch erneuerbare Energien in Deutschland im Jahr 2013—Eine erste Abschätzung," May 2014; Figure 1 also from earlier annual editions of this report.

31. Ibid.

32. "Krise Kostet Solarbranche fast 3000 Arbeitsplätze," *Die Welt*, 28 January 2014; "Photovoltaik-Produzenten verzeichnen Jobverluste," *pv magazine Deutschland*, 28 January 2014.

33. Chris Rose, "A Closer Look at ... Spain," *Wind Directions*, November 2013.

34. Asociación de Productores de Energía Renovables, *Study of the Macroeconomic Impact of Renewable Energies in Spain, Year 2012* (Madrid: 2013).

35. Ibid.

36. Ibid.

37. Ministério do Trabalho Emprego / Relação Anual de Informações Sociais, "Annual List of Social Information Database: Including Active and Inactive Employments for Sugarcane Cultivation and Alcohol Manufacture," online database; information provided by Renata Grisoli, researcher at Centro Nacional de Referência em Biomassa, São Paulo, discussion with Michael Renner, 5 March 2014.

38. Ministério de Minas e Energia and Empresa de Pesquisa Energética, "Plano Decenal de Expansão de Energie 2022," Brasilia, 2014, at www.epe.gov.br/PDEE/20140124_1.pdf.

39. Associação dos Produtores de Biodiesel do Brasil and Fundação Instituto de Pesquisas Econômicas, "Impactos Socioeconômicos da Indústria de Biodiesel no Brasil. Estudo Encomendado Pela Associação dos Produtores de

Biodiesel do Brasil," São Paulo, 2012, at www.aprobio.com .br/AprobioFIPERelatorioFinalsetembro2012.pdf.

40. Moana Simas and Sergio Pacca, "Energia Eólica, Geração de Empregos e Desenvolvimento Sustentável," *Estudos Avançados*, vol. 27, no. 77 (2013).

41. Figure 5 derived from Solar Foundation, *National Solar Jobs Census 2013: The Annual Review of the U.S. Solar Workforce* (Washington, DC: 2014) and earlier editions, from American Wind Energy Association (AWEA), *U.S. Wind Industry Annual Market Report 2013* (Washington, DC: 2014) and earlier editions, and from John M. Urbanchuk, "Contribution of the Ethanol Industry to the Economy of the United States," ABF Economics, Doylestown, PA, 17 February 2014 and earlier editions.

42. Solar Foundation, op. cit. note 41. This includes all solar technologies: PV, CSP, and solar heating/cooling.

43. Ibid.

44. Installations from Bloomberg New Energy Finance, *2014 Sustainable Energy in America Factbook* (Washington, DC: 2014); jobs from AWEA, op. cit. note 41.

45. Urbanchuk, op. cit. note 41.

46. Ibid.

47. Figure for 2003 from Nazmul Haque, Director and Head of Investment, Infrastructure Development Company Ltd., "IDCOL Solar Home System Program in Bangladesh," International Off-Grid Renewable Energy Conference, Accra, Ghana, 1–2 November 2012, at iorec.org/pdf/1_Session%203.pdf; 2013 from Dipal Chandra Barua, "Financial and Social Benefits of Building Energy Lending Program: Bangladesh Success Story," Zayed Future Energy Price presentation, Abu Dhabi, 20 January 2014.

48. Barua, op. cit. note 47.

49. Mika Ohbayashi, Japan Renewable Energy Foundation, e-mail to Michael Renner, 11 March 2014.

50. Masumi Suga and Chisaki Watanabe, "Japan Domestic Solar Shipments More Than Doubled in Fiscal 2013," *Bloomberg News*, 12 June 2014.

51. International Energy Agency, Co-operative Programme on PV Power Systems, *Photovoltaic Power Systems Programme Annual Report 2012* (Paris: 2013).

Chronic Hunger Falling, But One in Nine People Still Affected (pages 105–08)

1. U.N. Food and Agriculture Organization (FAO), *The State of Food and Agriculture* (Rome: 2013).

2. Ibid.

3. FAO, International Fund for Agricultural Development (IFAD), and World Food Programme (WFP), *The State of Food Insecurity in the World: Strengthening the Enabling Environment for Food Security and Nutrition* (Rome: FAO, 2014).

4. FAO, op. cit. note 1.

5. World Food Programme, "Hunger Glossary," at www.wfp .org/hunger/glossary.

6. FAO, op. cit. note 1.

7. UNICEF, *Improving Child Nutrition: The Achievable Imperative for Global Progress* (New York: 2013).

8. Ibid.

9. Ibid.

10. FAO, IFAD, and WFP, op. cit. note 3.

11. Ibid.

12. Ibid.

13. Ibid.

14. FAO, *FAO Statistical Yearbook 2014: Latin America and the Caribbean Food and Agriculture* (Santiago: 2014).

15. Ibid.

16. FAO, IFAD, and WFP, op. cit. note 3.

17. Ibid.

18. Ibid.

19. Ibid.

20. Ibid.

21. Ibid.

22. Ibid.

23. FAO, *Declaration of the World Summit on Food Security* (Rome: FAO, 2009); FAO, IFAD, and WFP, op. cit. note 3.

24. FAO, *Climate Change and Food Security: A Framework Document* (Rome: FAO, 2008).

25. Intergovernmental Panel on Climate Change, *Climate Change 2014: Impacts, Adaptation and Vulnerability* (Cambridge, U.K.: Cambridge University Press, 2014).

26. Ibid.

27. FAO, op. cit. note 14.

28. FAO, "World Food Situation: FAO Food Price Index," at www.fao.org/worldfoodsituation/foodpricesindex/en.

29. FAO, *FAO Statistical Yearbook 2014: Africa Food and Agriculture* (Accra: 2014).

30. FAO, IFAD, and WFP, op. cit. note 3.

31. FAO, op. cit. note 29.

32. FAO, IFAD, and WFP, op. cit. note 3.

33. Ibid.

34. World Bank, *Food Price Watch* (Washington, DC: 2014).

35. Ibid.

36. Ibid.

37. FAO, IFAD, and WFP, op. cit. note 3.

38. Ibid.

39. FAO, "FAO to Support Efforts Against Deepening Food Insecurity in Iraq," press release (Rome: 3 October 2014); FAO, IFAD, and WFP, op. cit. note 3.

40. Charles E. Hanrahan and Carol Canada, *International Food Aid: U.S. and Other Donor Contributions* (Washington, DC: Congressional Research Service, 2013).

41. Ibid.

42. Ibid.

43. Food Assistance Committee, *Commitments. Food Assistance Convention*, at www.foodassistanceconvention.org/en /commitments/commitments.aspx.

44. Jennifer Clapp, "The 2012 Food Assistance Convention: Is a Promise Still a Promise?" Centre for International Governance Innovation, 25 May 2012, at www.cigionline.org /blogs/inside-world-food-economy/2012-food-assistance -convention-promise-still-promise.

45. Ibid.

46. UNICEF, op. cit. note 7.

The Vital Signs Series

Some topics are included each year in *Vital Signs*; others are covered only in certain years. The following is a list of topics covered in *Vital Signs* thus far, with the year or years they appeared indicated in parentheses. The reference to 2006 indicates *Vital Signs 2006–2007*; 2007 refers to *Vital Signs 2007–2008*. The year 2013 indicates *Volume 20*; 2014 indicates *Volume 21*; and 2015 is this edition, *Volume 22*.

ENERGY AND TRANSPORTATION

Fossil Fuels
Carbon Use (1993)
Coal (1993–96, 1998, 2009, 2011, 2015)
Coal and Natural Gas Combined (2013)
Fossil Fuels Combined (1997, 1999–2003, 2005–07, 2010, 2014)
Natural Gas (1992, 1994–96, 1998, 2011–12)
Oil (1992–96, 1998, 2009, 2012–13)
Shale Gas (2015)

Renewables, Efficiency, Other Sources
Biofuels (2005–07, 2009–12, 2014)
Biomass Energy (1999)
Combined Heat and Power (2009)
Compact Fluorescent Lamps (1993–96, 1998–2000, 2002, 2009)
Efficiency (1992, 2002, 2006, 2015)
Geothermal Power (1993, 1997)
Hydroelectric Power (1993, 1998, 2006, 2012)
Hydropower and Geothermal Combined (2013)
Nuclear Power (1992–2003, 2005–07, 2009, 2011–12, 2014)
Smart Grid (2013, 2015)
Solar Power (1992–2002, 2005–07, 2009–12, 2015)
Solar and Wind Power (2014–15)
Solar Thermal Power (2010)
Wind Power (1992–2003, 2005–07, 2009–13, 2015)

Transportation
Air Travel (1993, 1999, 2005–07, 2011, 2014)
Bicycles (1992–2003, 2005–07, 2009)
Car-sharing (2002, 2006)
Electric Cars (1997)

Gas Prices (2001)
High-Speed Raid (2012)
Motorbikes (1998)
Railroads (2002, 2015)
Urban Transportation (1999, 2001)
Vehicles (1992–2003, 2005–07, 2009–15)

ENVIRONMENT AND CLIMATE

Atmosphere and Climate
Agriculture as Source of Greenhouse Gases (2014)
Carbon and Temperature Combined (2003, 2005–07, 2009–10)
Carbon Capture and Storage (2012–13)
Carbon Emissions (1992, 1994–2002, 2009, 2013–15)
CFC Production (1992–96, 1998, 2002)
Global Temperature (1992–2002)
Ozone Layer (1997, 2007)
Sea Level Rise (2003, 2011, 2015)
Weather-related Disasters (1996–2001, 2003, 2005–07, 2009–11, 2013–14)

Natural Resources, Animals, Plants
Amphibians (1995, 2000)
Aquatic Species (1996, 2002)
Birds (1992, 1994, 2001, 2003, 2006)
Coral Reefs (1994, 2001, 2006, 2010)
Dams (1995)
Ecosystem Conversion (1997)
Energy Productivity (1994, 2012)
Forests (1992, 1994–98, 2002, 2005–06, 2012)
Groundwater (2000, 2006)
Ice Melting (2000, 2005)
Invasive Species (2007)

Traffic Accidents (1994)
Tuberculosis (2000)
Military
Armed Forces (1997)
Arms Production (1997)
Arms Trade (1994)
Landmines (1996, 2002)
Military Expenditures (1992, 1998, 2003, 2005–06, 2014)
Nuclear Arsenal (1992–96, 1999, 2001, 2005, 2007)
Peacekeeping Expenditures (1994–2003, 2005–07, 2009, 2014)
Resource Wars (2003)
Small Arms (1998–99)
Wars (1995, 1998–2003, 2005–07)
Reproductive Health and Women's Status
Family Planning Access (1992)
Female Education (1998)
Fertility Rates (1993)
Gender Gap (2012)
Maternal Mortality (1992, 1997, 2003)
Population Growth (1992–2003, 2005–07, 2009–11, 2014–15)
Sperm Count (1999, 2007)
Violence Against Women (1996, 2002)
Women in Politics (1995, 2000, 2014)

Other Social Topics
Aging Populations (1997)
Climate Change Migration (2013)
Co-operatives (2013)
Educational Levels (2011)
Homelessness (1995)
Income Distribution or Poverty (1992, 1995, 1997, 2002–03, 2010)
International Criminal Court (2003)
Language Extinction (1997, 2001, 2006)
Literacy (1993, 2001, 2007)
Millennium Development Goals (2005, 2007)
Nongovernmental Organizations (1999)
Orphans Due to AIDS Deaths (2003)
Prison Populations (2000)
Public Policy Networks (2005)
Quality of Life (2006)
Refugees (1993–2000, 2001, 2003, 2005, 2014)
Refugees-Environmental (2009)
Religious Environmentalism (2001)
Slums (2006)
Social Security (2001)
Sustainable Communities (2007)
Teacher Supply (2002)
Urbanization (1995–96, 1998, 2000, 2002, 2007, 2013)
Voter Turnouts (1996, 2002)

Island Press | Board of Directors